Mobile Computing & Wireless Communication

From Basics to Essentials

MOHIT THAKKAR

About the Book:

It often happens that when we try to study a subject for some examination or a job interview, we just don't find the right content.

The problem with the reference books is that they are too descriptive for last moment studies. Whereas the problem with local publications is that they are inaccurate as compared to the reference books.

This particular book encapsulates the subject notes on **Mobile Computing & Wireless Communication** with the combined benefits of reference books & local publications. It has the accuracy of a reference book as well as the abstraction of a local publication.

The author studied the subject from various sources such as web lectures, reference books, online tutorials & so on. After having a thorough understanding of the subject, the author compiled this book for an easy understanding of the subject.

This book presents the content with utmost simplicity of language, and in an abstract manner so that it can be used for last moment studies. This book can be used by:

- Students to prepare for their examinations

- Professionals to prepare for job interviews.

- Individuals willing to have a basic understanding of the domain: **Mobile Computing & Wireless Communication**.

Happy Reading!

References:

1. Data Communications and Networking, 4th Edition, by Behrouz A. Forouzan, McGraw-Hill Higher Education

2. Wireless Communications & Networks, 2nd Edition, by William Stallings, Pearson Education

3. Android Wireless Application Development, 3rd Edition, by Lauren Darcey & Shane Conder, Addison Wesley

Disclaimer:

The content in this book is compiled from various sources & also contains personal views of the author. Some of the content in this book might be inherited from other books or web resources. The author claims no rights to such content. The rights to any such content remains with the respective owners (mentioned in the References).

© 2018 Mohit Thakkar

All rights reserved. No part of this publication may be reproduced, stored in a retrieval system, or transmitted in any form or by any means, electronic, mechanical, photocopying, recording, or otherwise, without prior written permission of the publisher.

Contents:

- About the Book i
- References ii
- Disclaimer ii
- Contents iii
- List of Figures vii

- **Chapter 1: Introduction to Transmission Fundamentals** 1
    - Transmission Fundamentals 2
        - Signals for Conveying Information 2
        - Analog and Digital Data Transmission 6
        - Channel Capacity 9
        - Transmission Media 12
        - Multiplexing 18
    - Communication Networks 20
        - LANs, MANs, and WANs 20
        - Switching Techniques 22
        - Circuit Switching 23
        - Packet Switching 24
    - Protocols and the TCP/IP Suite 27
        - The Need for a Protocol Architecture 27
        - The TCP/IP Protocol Architecture 28
        - The OSI Model 30
        - OSI versus TCP/IP 33
        - Internetworking 34

- **Chapter 2: Introduction to Mobile Computing** 35
    - Mobile Computing 36
        - Introduction to Mobile Computing & it's Principles 36
        - Architecture of Mobile Computing (Three Tier) 37
        - Advantages of Mobile Computing 39
        - Security issues in Mobile Computing 40
    - Cellular Wireless Networks 42
        - Introduction to Cellular Networks 42
        - Principles of Cellular Network 44
        - Frequency Reuse in a Cellular Network 45

- Handoff/Handover in Cellular Networks 46
- 1G: First Generation Networks 47
- 2G: Second Generation Networks 48
- 2.5G Mobile Networks 49
- 3G: Third Generation Networks 50
- 1G, 2G, 2.5G, 3G - Comparison 51
- Antennas and Propagation 52
  - Antennas 52
  - Propagation Modes 56
  - Line of Sight Transmission 59
  - Fading in the Mobile Environment 62
- Spread Spectrum 64
  - The Concept of Spread Spectrum 64
  - Frequency Hopping Spread Spectrum (FHSS) 65
  - Direct Sequence Spread Spectrum (DSSS) 67
- Coding and Error Control 69
  - Error Detection 69
  - Block Error Correction Code 71
  - Automatic Repeat Request (ARQ) 73

➤ Chapter 3: Introduction to GSM & GPRS 75

- Multiple access in Wireless System 76
  - Multiple Access Scheme 76
  - Frequency Division Multiple Access (FDMA) 77
  - Time Division Multiple Access (TDMA) 78
  - Code Division Multiple Access (CDMA) 79
  - Space Division Multiple Access (SDMA) 81
- Global System for Mobile Communication (GSM) 82
  - Introduction to GSM 82
  - GSM Architecture 83
  - Frequency Reuse in GSM 86
  - Roaming in GSM 87
  - Handover in GSM 88
  - Call Routing in GSM 90
  - GSM Services 93
  - GSM versus CDMA 95
- General Packet Radio Service (GPRS) 96
  - Introduction to GPRS 96
  - GPRS Architecture 97

- PDP Context Activation in GPRS 99
- Routing in GPRS 101
- Data Services in GPRS 103
- Billing & Charging in GPRS 104
- Applications & Limitations of GPRS 105
- Mobile IP 107
  - Working of Mobile IP (Architecture)... 107
  - Discovery, Registration & Tunneling 109
  - Traditional IP versus Mobile IP 111

- **Chapter 4: Introduction to Wifi** 113
  - Wireless LAN (Wi-Fi) - Advantages, Disadvantages & Goals 114
  - Types of Wireless LANs – Adhoc Mode versus Infrastructure Mode ...116
  - IEEE 802 Architecture...117
  - IEEE 802.11 Architecture & Services..120
  - IEEE 802.11 Medium Access Control 123
  - Wi-Fi Protected Access 124
  - 3G versus Wi-Fi 126

- **Chapter 5: Introduction to Bluetooth** 127
  - Bluetooth Applications... 128
  - Piconet & Scatternet in Bluetooth 130
  - Bluetooth Architecture (Protocol Stack).. 132

- **Chapter 6: Introduction to Android**... 135
  - Android Architecture & Application Framework 136
  - The Manifest File 140
  - Android Layouts 143
    - The LinearLayout 143
    - The RelativeLayout 144
    - The ScrollView Layout 145
    - The TableLayout 146
    - The FrameLayout 148
  - Using the TextView.. 149
  - EditText View 151
  - Button View 153
  - RadioButton 155

- CheckBox 157
- ImageButton 159
- RatingBar 161
- The ProgressBar 163
- The Context Menu 166

List of Figures:

- Figure 1.1: Analog & Digital Signals 2
- Figure 1.2: Periodic Signals..3
- Figure 1.3: Real-Time Signal 4
- Figure 1.4: Analog & Digital Signaling of Data 7
- Figure 1.5: Effect of Noise on a Digital Signal 10
- Figure 1.6: Classification of Transmission Media 12
- Figure 1.7: Twisted Pair Cable 13
- Figure 1.8: Unshielded Twisted Pair Cable 13
- Figure 1.9: Coaxial Cable 14
- Figure 1.10: Fiber Optic Cable 14
- Figure 1.11: Radio Wave Transmission 15
- Figure 1.12: Light Wave Transmission 17
- Figure 1.13: Multiplexing 18
- Figure 1.14: Frequency Division Multiplexing 19
- Figure 1.15: Time Division Multiplexing 19
- Figure 1.16: Local Area Network 20
- Figure 1.17: Metropolitan Area Network 21
- Figure 1.18: Wide Area Network 21
- Figure 1.19: Simple Switching Network 22
- Figure 1.20: Packet Switching (Datagram) 25
- Figure 1.21: Packet Switching (Virtual Circuit) 26
- Figure 1.22: TCP/IP Architecture 29
- Figure 1.23: OSI Architecture...31
- Figure 2.1: Architecture of Mobile Computing 37
- Figure 2.2: 3-Tier Architecture of Mobile Computing 38
- Figure 2.3: Cellular Network - 7 Cell Pattern 42
- Figure 2.4: Cellular Network – Cluster 43
- Figure 2.5: Frequency Reuse in Cellular Network 45
- Figure 2.6: Handoff in a Cellular Network 46
- Figure 2.7: Evolution of Cellular Wireless Networks...51
- Figure 2.8: Radiation Pattern of an Isotropic Antenna.......................................53
- Figure 2.9: Dipole Antennas 53
- Figure 2.10: Parabolic Reflective Antennas 54
- Figure 2.11: Ground Wave Propagation 56
- Figure 2.12: Sky Wave Propagation 57
- Figure 2.13: Line-of-Sight (LOS) Propagation..58
- Figure 2.14: Multipath Interference 61

- ➢ Figure 2.15: Propagation Mechanisms: Reflection, Scattering, Diffraction 62
- ➢ Figure 2.16: Spectral Communication in Wireless Networks 64
- ➢ Figure 2.17: Frequency Hopping Example 65
- ➢ Figure 2.18: Frequency Hopping Spread Spectrum System (Block Diagram 66
- ➢ Figure 2.19: Direct Sequence Spread Spectrum System 67
- ➢ Figure 2.20: Direct Sequence Spread Spectrum System (Block Diagram) 68
- ➢ Figure 2.21: Error Detection Process 70
- ➢ Figure 2.22: CRC Error Detection 70
- ➢ Figure 2.23: Forward Error Correction 71
- ➢ Figure 3.1: Frequency Division Multiple Access (FDMA) 76
- ➢ Figure 3.2: Time Division Multiple Access (TDMA) 77
- ➢ Figure 3.3: Code Division Multiple Access(CDMA) 78
- ➢ Figure 3.4: Example of CDMA 79
- ➢ Figure 3.5: Space Division Multiple Access 80
- ➢ Figure 3.6: GSM Architecture 82
- ➢ Figure 3.7: GSM System Hierarchy 83
- ➢ Figure 3.8: Cell Clusters in GSM 85
- ➢ Figure 3.9: Handoff in GSM 88
- ➢ Figure 3.10: Call Routing - Block Diagram 89
- ➢ Figure 3.11: Call Routing in GSM 90
- ➢ Figure 3.12: GPRS Architecture 96
- ➢ Figure 3.13: PDP Context Activation 98
- ➢ Figure 3.14: Routing in GPRS 100
- ➢ Figure 3.15: Architecture of Mobile IP 106
- ➢ Figure 4.1: Adhoc Mode versus Infrastructure Mode 114
- ➢ Figure 4.2: IEEE 802 Architecture in Reference to OSI model 115
- ➢ Figure 4.3: IEEE 802.11 Architecture 118
- ➢ Figure 4.4: IEEE 802.11 Services 120
- ➢ Figure 4.5: Wi-Fi Protected Access (WPA) 122
- ➢ Figure 5.1: Master/Slave Relationships in Scatternet 128
- ➢ Figure 5.2: Scatternet 129
- ➢ Figure 5.3: Wireless Network Configurations 129
- ➢ Figure 5.4: Bluetooth Protocol Stack 131
- ➢ Figure 6.1: Android Architecture 134
- ➢ Figure 6.2: Linear Layout – Android 141
- ➢ Figure 6.3: Relative Layout – Android 142
- ➢ Figure 6.4: Table Layout – Android 144
- ➢ Figure 6.5: Frame Layout – Android 146
- ➢ Figure 6.6: TextView – Android 148
- ➢ Figure 6.7: EditText View – Android 150

- ➢ Figure 6.8: EditText View - Styles – Android 150
- ➢ Figure 6.9: Button in Android 151
- ➢ Figure 6.10: RadioButton – Android 153
- ➢ Figure 6.11: Checkbox – Android 156
- ➢ Figure 6.12: ImageButton – Styles – Android 157
- ➢ Figure 6.13: ImageButton – Android 158
- ➢ Figure 6.14: RatingBar – Android 160
- ➢ Figure 6.15: ProgressBar – Android 161
- ➢ Figure 6.16: Context Menu – Android 166

Introduction to Transmission Fundamentals

1.1 Transmission Fundamentals:

1.1.1 Signals for Conveying Information:

Here, we are concerned with **electromagnetic signals** used as a means to **transmit information**. An electromagnetic signal is a function of time, but it can also be expressed as a function of frequency; that is, the signal consists of components of different frequencies. It turns out that the frequency domain view of a signal is far more important to an understanding of data transmission than a time domain view.

Time Domain Concepts:

As a function of time, an electromagnetic signal can be either **analog** or **digital**.

An **analog signal** is one in which the signal **intensity varies in a smooth fashion** over time. In other words, there are no breaks or discontinuities in the signal. Such signals might represent speech.

A **digital signal** is one in which the signal **intensity maintains a constant level for some period of time and then changes to another constant level**. Such signals might represent binary 0s & 1s.

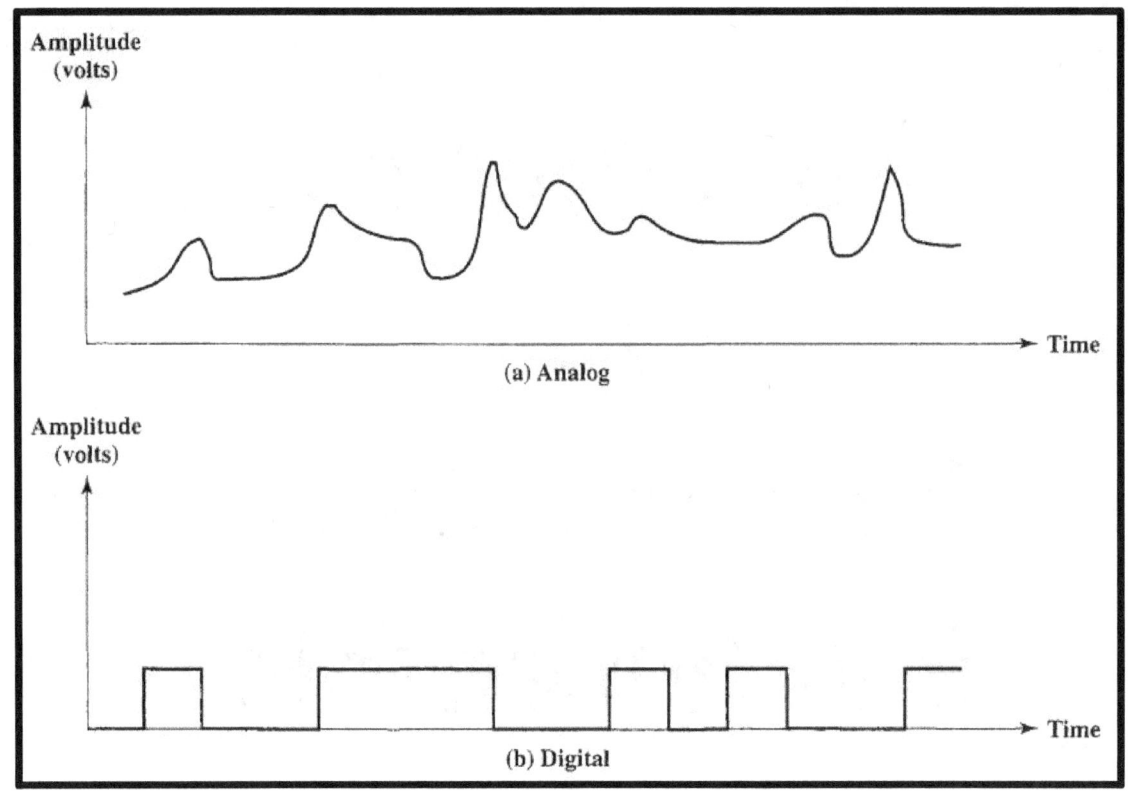

Figure 1.1: Analog & Digital Signals

A *periodic signal* is the one in which the same signal pattern repeats over time. Figure 2 shows an example of a *periodic analog signal (sine wave)* and *a periodic digital signal (square wave)*.

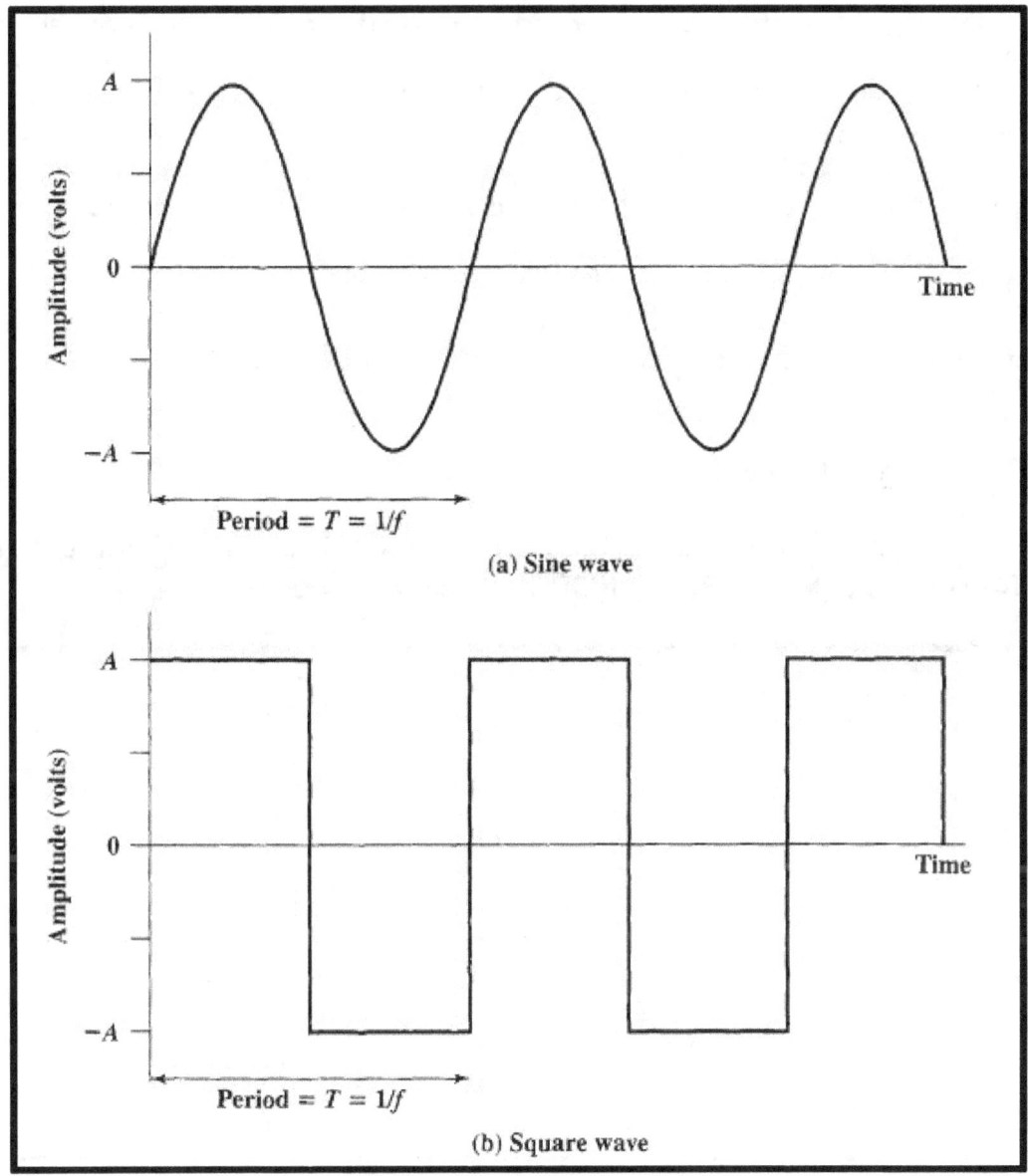

Figure 1.2: Periodic Signals

Mathematically, a signal set s(t) is defined to be periodic if and only if

$$s(t + T) = s(t) \quad \{-\infty < t < +\infty\}$$

where the constant T is the period of the signal (T is the smallest value that satisfies the equation). Otherwise, a signal is *aperiodic*. The *sine wave* is the *fundamental analog signal* & can be represented by three parameters: *Peak Amplitude (A)*, *Frequency (f)*, and *Phase (φ)*.

The **Peak Amplitude (A)** is the maximum value or strength of the signal over time; typically, this value is measured in **volts**.

The **Frequency (f)** is the rate [in **cycles per second**, or **Hertz (Hz)**] at which the signal repeats.

An equivalent parameter is the **Period (T)** of a signal, which is the amount of time it takes for one repetition; therefore, **T = 1/f**.

Phase (ϕ) is a measure of the relative position in time within a single period of a signal.

The general sine wave can be written as: **s(t) = A sin(2πft + ϕ)**

The W**avelength (λ)** of a signal is the distance occupied by a single cycle.

Frequency Domain Concepts:

In practice, an electromagnetic signal will be made up of many frequencies. For example, the signal **s(t) = (4/π) x (sin(2πft) + (1/3)sin(2π(3f)t))** is shown in the below figure:

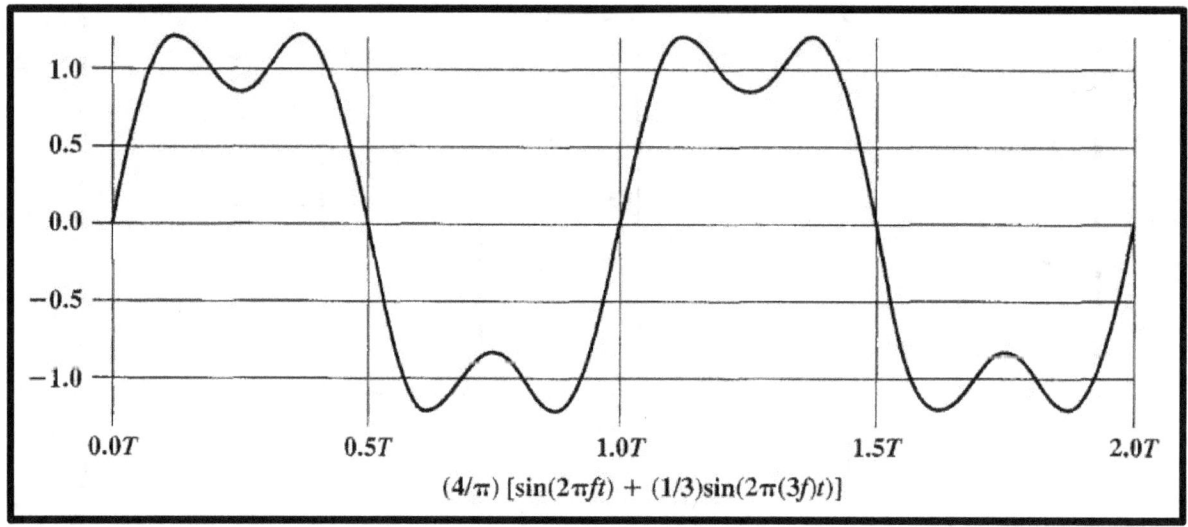

Figure 1.3: Real-Time Signal

The components of this signal are just sine waves of frequencies f and 3f. There are two interesting points that can be made about this figure:

- The second frequency is an integer multiple of the first frequency. When all of the frequency components of a signal are integer multiples of one frequency, the latter frequency is referred to as the ***fundamental frequency***.
- The period of the total signal is equal to the period of the fundamental frequency. The period of the component sin(2πft) is T = 1/f, and the period of s(t) is also T.

Using a discipline known as Fourier analysis, any electromagnetic signal can be shown to consist of a collection of periodic analog signals (sine waves) at different amplitudes, frequencies, and phases.

The **Spectrum** of a signal is the range of frequencies that it contains. For the signal of figure 3, the spectrum extends from f to 3f.

The **Absolute Bandwidth** of a signal is the width of the spectrum. In the case of Figure 3, the bandwidth is 3f - f = 2f.

Many signals have an infinite bandwidth, but with most of the energy contained in a relatively narrow band of frequencies. This band is referred to as the **Effective Bandwidth**, or just **Bandwidth**.

Relationship between Data Rate and Bandwidth

- The greater the bandwidth, the higher the information-carrying capacity.
- Any digital waveform will have infinite bandwidth **BUT** the transmission system will limit the bandwidth that can be transmitted.
- For any given medium, the greater the bandwidth transmitted, the greater the cost.
- **HOWEVER**, limiting the bandwidth creates distortions.

1.1.2 Analog and Digital Data Transmission:

The terms analog and digital correspond, roughly, to continuous and discrete, respectively. These two terms are used frequently in data communications in at least three contexts: **Data**, **Signals**, and **Transmission**.

- **Data** - entities that convey meaning, or information
- **Signals** - electric or electromagnetic representations of data
- **Transmission** - communication of data by the propagation and processing of signals

1.2.1 Analog & Digital Data

Analog Data take on continuous values in some interval. For example, voice and video are continuously varying patterns of intensity. Most data collected by sensors, such as temperature and pressure, are continuous valued.

Digital Data take on discrete values; examples are text and integers.

1.2.2 Analog and Digital Signaling

An **analog signal** is a continuously varying electromagnetic wave that may be propagated over a variety of media, depending on frequency; examples are **copper wire media**, such as twisted pair and coaxial cable; **fiber optic cable**; and **atmosphere or space propagation** (wireless).

A **digital signal** is a sequence of discrete voltage pulses that may be transmitted over a **copper wire medium**; for example, a **constant positive voltage** level may represent **binary 0** and a **constant negative voltage** level may represent **binary 1**.

Analog Signal	Digital Signal
Continuously varying electromagnetic wave	Sequence of discrete voltage pulses
Expensive	Relatively cheaper
More susceptible to noise interference	Less susceptible to noise interference
Suffer less from attenuation (reduction of signal strength at higher frequencies)	Suffer more from attenuation (reduction of signal strength at higher frequencies)

1.2.3 Possible Combinations of Data & Signals

Both analog and digital data can be represented and propagated by either analog or digital signals. The following figure shows four different possibilities:

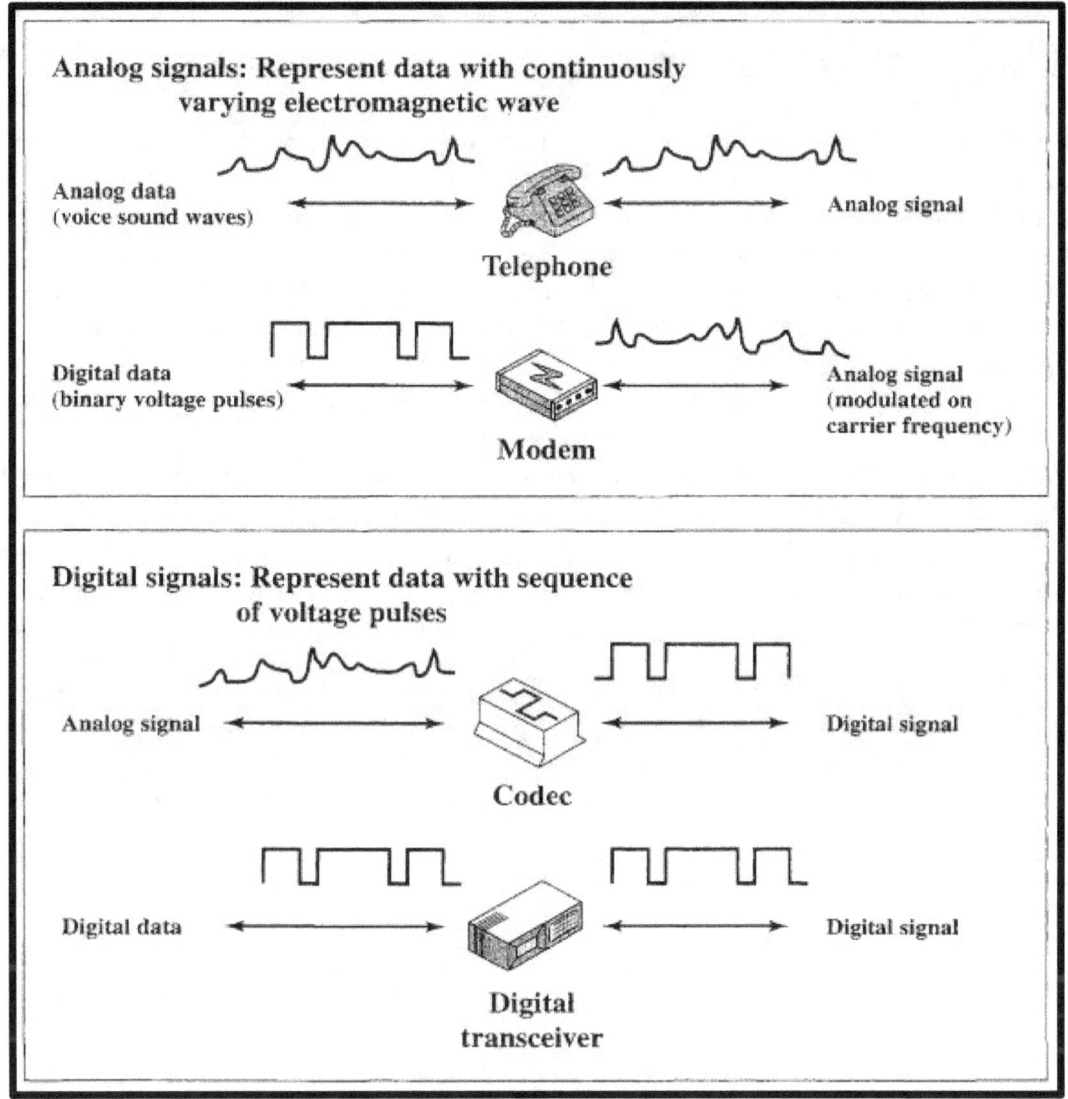

Figure 1.4: Analog & Digital Signaling of Data

Following are the reasons to choose each of the possibilities:

A. **_Digital data, Digital signal_**: In general, the equipment for encoding digital data into a digital signal is less complex and less expensive than digital to analog equipment.

B. **_Analog data, Digital signal_**: Conversion of analog data to digital form permits the use of modern digital transmission and switching equipment for analog data.

C. **_Digital data, Analog signal_**: Some transmission media, such as optical fiber and satellite, will only propagate analog signals.

D. **_Analog data, Analog signal_**: Analog data are easily converted to an analog signal.

1.2.4 Analog and Digital Transmission

Analog transmission is a means of transmitting analog signals **without regard to their content**; the signals may represent **analog data** (e.g., voice) or **digital data** (e.g., data that pass through a modem), **in either case, the signal will suffer attenuation** (reduction of signal strength at higher frequencies). To achieve longer distances, the analog transmission system includes **amplifiers** that **boost the energy in the signal**. Unfortunately, the amplifier **also boosts the noise components**. With amplifiers cascaded to achieve long distance, the **signal becomes more and more distorted**. For analog data, such as voice, quite a bit of distortion can be tolerated and the data remain intelligible. However, for digital data transmitted as analog signals, cascaded amplifiers will introduce errors.

Digital transmission, in contrast, is **concerned with the content** of the signal. We have mentioned that a digital signal can be propagated only for a limited distance before attenuation endangers the integrity of the data. **To achieve greater distances, repeaters are used**. A *repeater* receives the digital signal, recovers the pattern of ones and zeros, and **retransmits a new signal** to overcome attenuation.

1.1.3 Channel Capacity:

The maximum rate at which data can be transmitted over a given channel, under given conditions is referred to as the **Channel Capacity.**

There are four concepts here that we are trying to relate to one another:
- **Data rate**: This is the rate, in bits per second (bps), at which data can be communicated.
- **Bandwidth**: This is the bandwidth of the transmitted signal as constrained by the transmitter and the nature of the transmission medium, expressed in cycles per second, or Hertz.
- **Noise**: For this discussion, we are concerned with the average level of noise over the communications path.
- **Error rate**: This is the rate at which errors occur, where an error is the reception of a 1 when a 0 was transmitted or the reception of a 0 when a 1 was transmitted.

The problem we are addressing is that communications facilities are expensive and **the greater the bandwidth of a channel, the greater the cost**.

Following portion will cover the two different formulas to calculate a channel's capacity.

Nyquist Bandwidth

Let us consider the case of a **channel that is noise free**. In this case, the limitation on data rate is simply the bandwidth of the signal.

Nyquist formulation for this scenario states that **if the rate of signal transmission is 2B, then a signal with frequencies of B is sufficient to carry the signal rate**.

The converse is also true: Given a bandwidth of B, the highest signal rate that can be carried is 2B.

With multilevel signaling, the **Nyquist formulation** becomes $C = 2B \log_2 M$
Where,   **C: Channel Capacity**
         B: Bandwidth of the channel
         M: number of discrete signal elements or voltage levels

Generally, in a **digital signal**, we have **two voltage levels: 0 & 1**. In such cases, **M will be 2** & $\log_2 M$ **will evaluate to 1**.

Thus, for M = 8, a bandwidth of B = 3100 Hz yields a capacity C = 18,600 bps.

Shannon Bandwidth

Nyquist's formula indicates that ***doubling the bandwidth doubles the data rate***. Now, consider the relationship among ***data rate***, ***noise***, and ***error rate***.

If the data rate is increased, then the number of bits per second becomes higher, so that more bits are affected by a given pattern of noise. Thus, ***the higher the data rate, the higher the error rate***. The following figure shows the effect of noise on a signal:

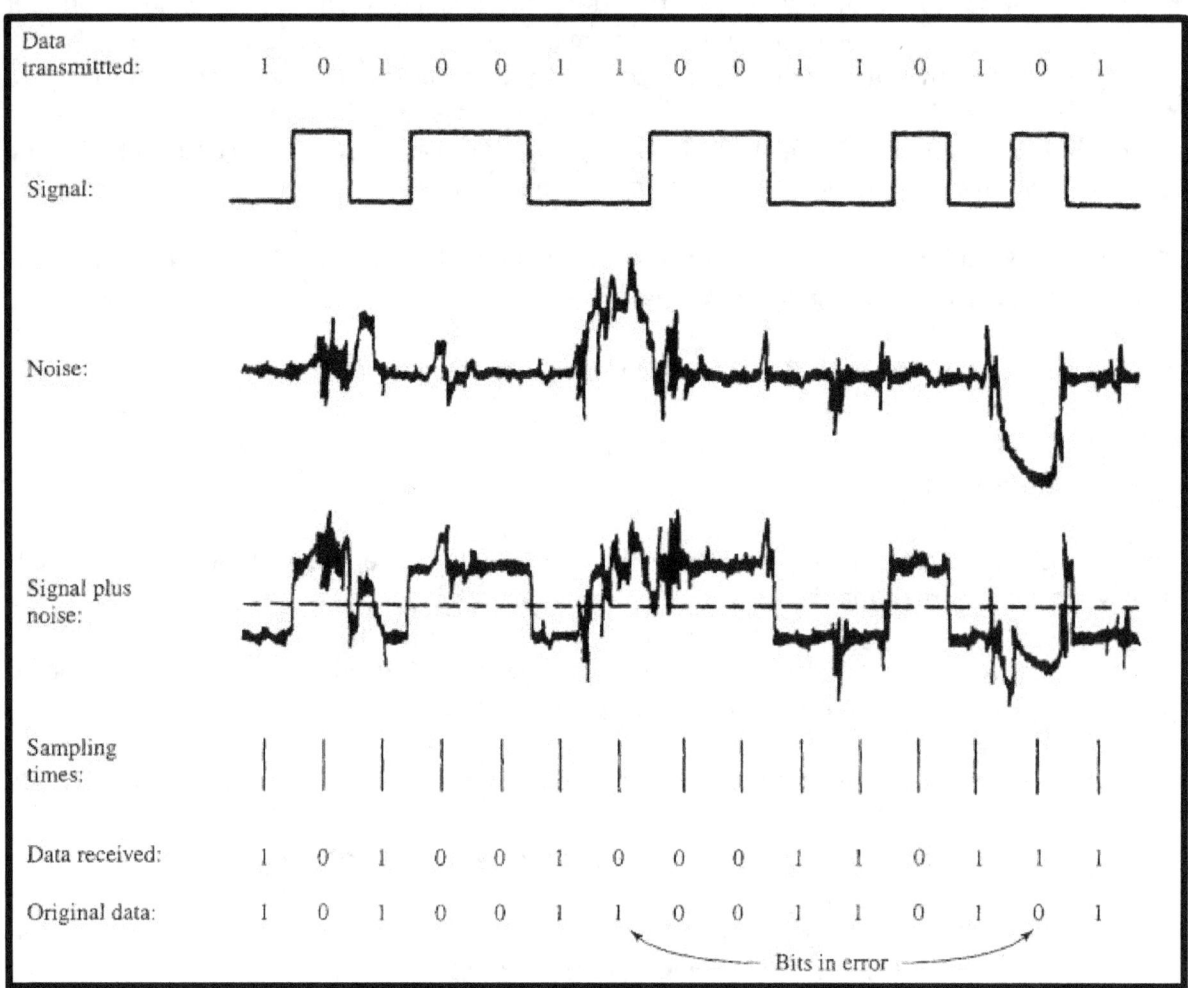

Figure 1.5: Effect of Noise on a Digital Signal

For a given level of noise, we would expect that a ***greater signal strength*** would improve the ability to ***receive data correctly in the presence of noise***. The key parameter involved in this reasoning is the ***signal-to-noise ratio (SNR)***, which is ratio of the power in a signal to the power in the noise. Typically, this ratio is measured at a receiver, because it is the point where an attempt is made to eliminate the unwanted noise from the signal. ***SNR is measured in Decibels***.

$$SNR_{DB} = 10 \log_{10} \frac{\text{signal power}}{\text{noise power}}$$

Shannon's formula for **maximum channel capacity**, in **bits per second**, is as follows:

C = B log$_2$(1 + SNR)

Where, C: *Channel Capacity*
B: *Bandwidth of the channel in Hertz*
SNR: *Signal to Noise ratio*

Example : Let us consider an example that relates the Nyquist and Shannon formulations. Suppose that the spectrum of a channel is between 3 MHz and 4 MHz and SNR_{dB} = 24 dB. Then

$$B = 4\,\text{MHz} - 3\,\text{MHz} = 1\,\text{MHz}$$
$$SNR_{dB} = 24\,\text{dB} = 10\log_{10}(SNR)$$
$$SNR = 251$$

Using Shannon's formula,

$$C = 10^6 \times \log_2(1 + 251) \approx 10^6 \times 8 = 8\,\text{Mbps}$$

This is a theoretical limit and, as we have said, is unlikely to be reached. But assume we can achieve the limit. Based on Nyquist's formula, how many signaling levels are required? We have

$$C = 2B \log_2 M$$
$$8 \times 10^6 = 2 \times (10^6) \times \log_2 M$$
$$4 = \log_2 M$$
$$M = 16$$

1.1.4 Transmission Media:

The **transmission medium** is the **physical path** between transmitter and receiver.

Transmission media can be classified as **guided** or **unguided**. In both cases, communication is in the form of electromagnetic waves.

Guided Media

With **guided media**, the waves are guided along a **solid medium**, such as copper twisted pair, copper coaxial cable, or optical fiber.

Unguided Media

The **atmosphere** and **outer space** are examples of **unguided** media, which provide a means of transmitting electromagnetic signals but do not guide them; this form of transmission is usually referred to as **wireless transmission**.

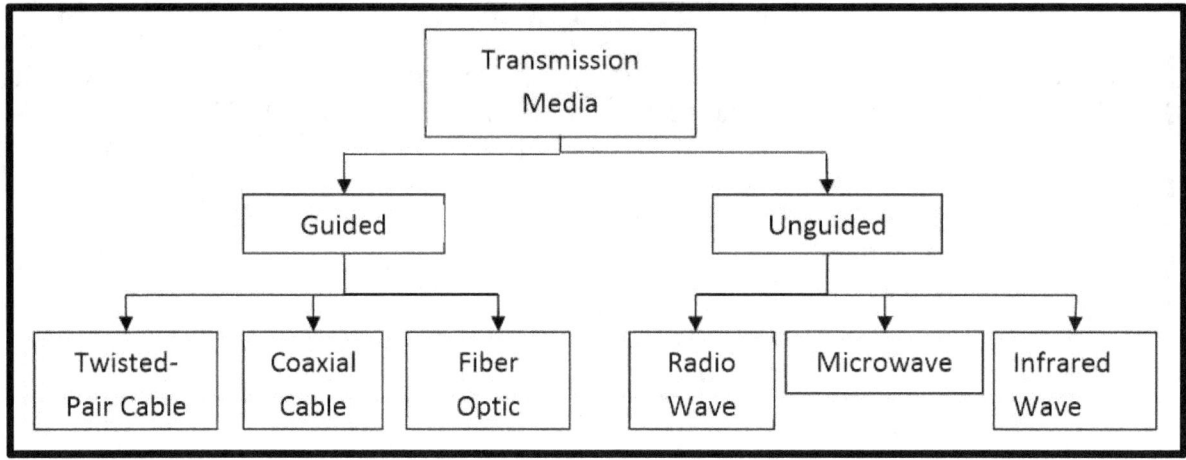

Figure 1.6: Classification of Transmission Media

Magnetic Media

One of the most common ways to transport data from one computer to another is to write them onto **magnetic tape** or **removable media** (e.g., recordable DVDs), physically transport the tape or disks to the destination machine, and read them back in again.

Although this method is not as sophisticated as using a geosynchronous communication satellite, it is often more **cost effective**, especially for applications in which high bandwidth or cost per bit transported is the key factor.

Twisted Pair Cable

A twisted pair consists of **two insulated copper wires**, typically about 1 mm thick. The wires are **twisted together** in a helical form, just like a DNA molecule.

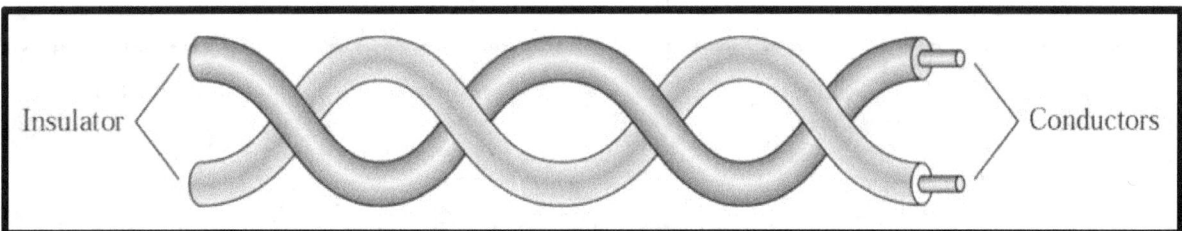

Figure 1.7: Twisted Pair Cable

Why cable is twisted?
If the two wires are parallel, the effect of these **unwanted signals** is not the same in both wires because they are at different locations relatives to the **noise or crosstalk** sources. This results in a difference at the receiver. By twisting the pair, a **balance is maintained**.

Types of Twisted-Pair Cable:

Unshielded twisted-pair (UTP)

Twisted pair cabling comes in several varieties, two of which are important for computer networks.

Figure 1.8: Unshielded Twisted Pair Cable

Category 3 twisted pairs consist of two insulated wires gently twisted together. Most office buildings had one category 3 cable running from a central wiring closet on each floor into each office.

Category 5 is the more advanced twisted pairs that were introduced. They are similar to category 3 pairs, but with **more twists per centimeter**, which results in **less crosstalk and a better-quality signal** over longer distances, making them more suitable for high-speed computer communication.

Shielded twisted-pair (STP)

STP cable has a metal foil or **braided mesh covering that encases each pair of insulated conductors**. Metal casing improves the quality of cable by **preventing the penetration of noise or crosstalk**. It is **bulkier** and **more expensive**. It's Used in telephone lines to provide voice and data channels.

Coaxial Cable

A *coaxial cable* consists of a *stiff copper wire as the core*, surrounded by an *insulating material*. The insulator is encased by a *cylindrical conductor*, often as a closely-woven braided mesh. The outer conductor is covered in a *protective plastic sheath*.

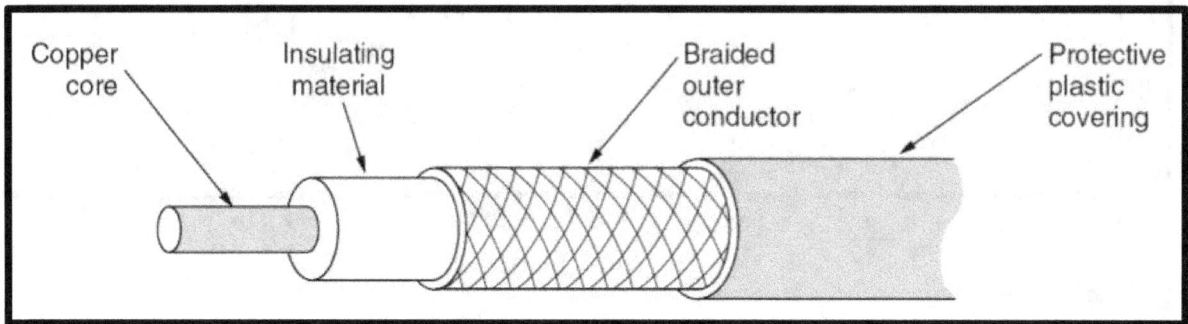

Figure 1.9: Coaxial Cable

Coaxial Cable has *better shielding* than twisted pairs, so it can *span longer distances* at *higher speeds*. The construction and shielding of the coaxial cable give it a *good combination of high bandwidth and excellent noise immunity*. *Modern coaxial cables* have a bandwidth of close to *1 GHz*.

Fiber Optics

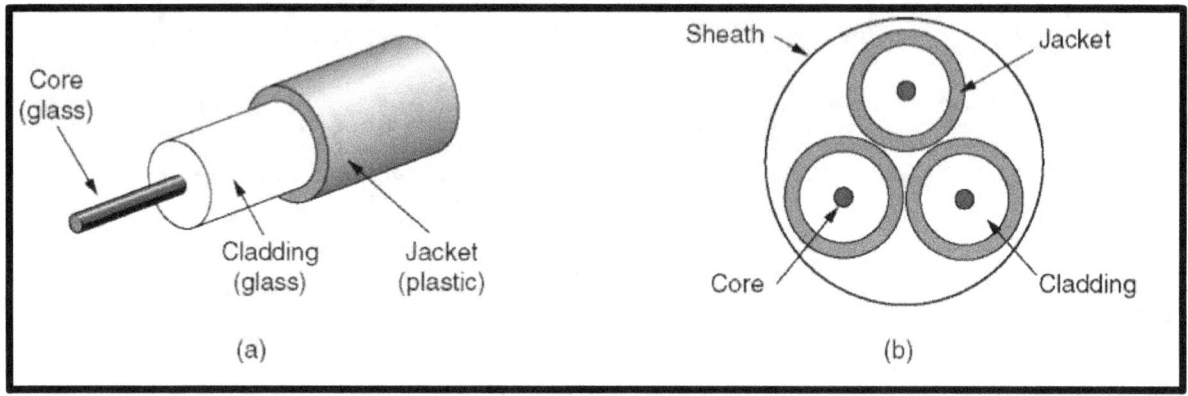

Figure 1.10: Fiber Optic Cable

A *fiber-optic cable* is made of *glass or plastic* and transmits *signals in the form of light.* Optical fibers use *reflection to guide light* through a channel. A *glass or plastic core* is surrounded by a *cladding of less dense glass or plastic*. The *difference in density* of the two materials must be such that a beam of *light moving through a core is reflected off the cladding* instead of being refracted into it.

Fiber optic cables are *similar to coaxial cables*, except *without the braid*. The above figure shows a single fiber viewed from the side. At the center is the glass core through which the light propagates.

The core is surrounded by a glass cladding with a lower index of refraction than the core, to keep all the light in the core. Next comes a thin plastic jacket to protect the cladding. Fibers are typically grouped in bundles, protected by an outer sheath. Figure shows a sheath with three fibers.

Radio Transmission

Radio waves are easy to generate, can travel long distances, and can penetrate buildings easily, so they are widely used for communication, both indoors and outdoors.

Radio waves also are *omnidirectional*, meaning that they travel in all directions from the source, so the *transmitter and receiver do not have to be carefully aligned* physically.

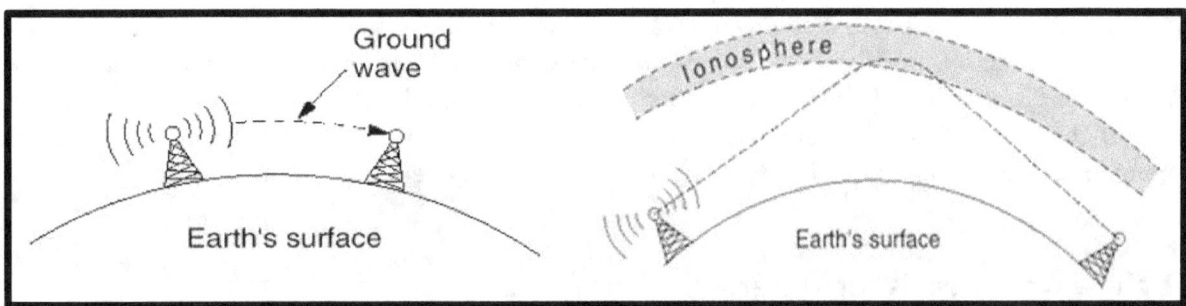

Figure 1.11: Radio Wave Transmission

The *properties of radio waves* are *frequency dependent*. At *low frequencies*, radio waves pass through obstacles well, but the power falls off sharply with distance from the source. At *high frequencies*, radio waves tend to travel in straight lines and bounce off obstacles. They are also absorbed by rain. At all frequencies, radio waves are subject to interference from motors and other electrical equipment. In the VLF, LF, and MF bands, radio waves follow the curvature of the earth. In the HF, they bounce off the ionosphere.

Microwave Transmission

Since the *microwaves travel in a straight line*, if the *towers are too far* apart, the *earth will get in the way*. Consequently, *repeaters are needed periodically.*

Unlike radio waves at lower frequencies, *microwaves do not pass through buildings well*. In addition, even if the beam is well focused at the transmitter, there is still some *divergence in space*. Concentrating all the energy into a small beam using a parabolic antenna gives a much higher signal to noise ratio.

Advantages: No right way is needed (compared to wired media).
Relatively inexpensive.
Simple to install.

Disavantages: Do not pass through buildings well.
Multipath fading problem (the delayed waves cancel the signal).
Absorption by rain above 8 GHz.
Severe shortage of spectrum.

Infrared Transmission

Unguided infrared waves are widely used for **short-range communication**. The **remote controls** used on televisions, VCRs, and stereos all **use infrared communication**.

They are relatively **directional, cheap, and easy to build** but have a major drawback: they **do not pass through solid objects**.

On the other hand, the fact that infrared waves do not pass through solid walls well *is also a plus*. It means that an **infrared system in one room of a building will not interfere with a similar system in adjacent rooms or buildings**. Furthermore, security of infrared systems against eavesdropping is better than that of radio systems precisely for this reason. Therefore, **no government license is needed to operate an infrared system**, in contrast to radio systems, which must be licensed.

Light wave Transmission (Laser)

A modern application is to connect the LANs in two buildings via lasers mounted on their rooftops. So, each building needs its own laser and its own photo detector.

Lasers are **unidirectional** with **high bandwidth** and **low cost**. It is also **easy to install** and does not require an FCC (Federal Communications Commission) license.

A **disadvantage** is that laser beams **can't penetrate rain or thick fog**, but they normally work well on sunny days.

Advantages: No right way is needed (compared to wired media).
Relatively inexpensive.
Simple to install.

Disadvantages: Do not pass through buildings well.
Multipath fading problem (the delayed waves cancel the signal).
Absorption by rain above 8 GHz.
Severe shortage of spectrum

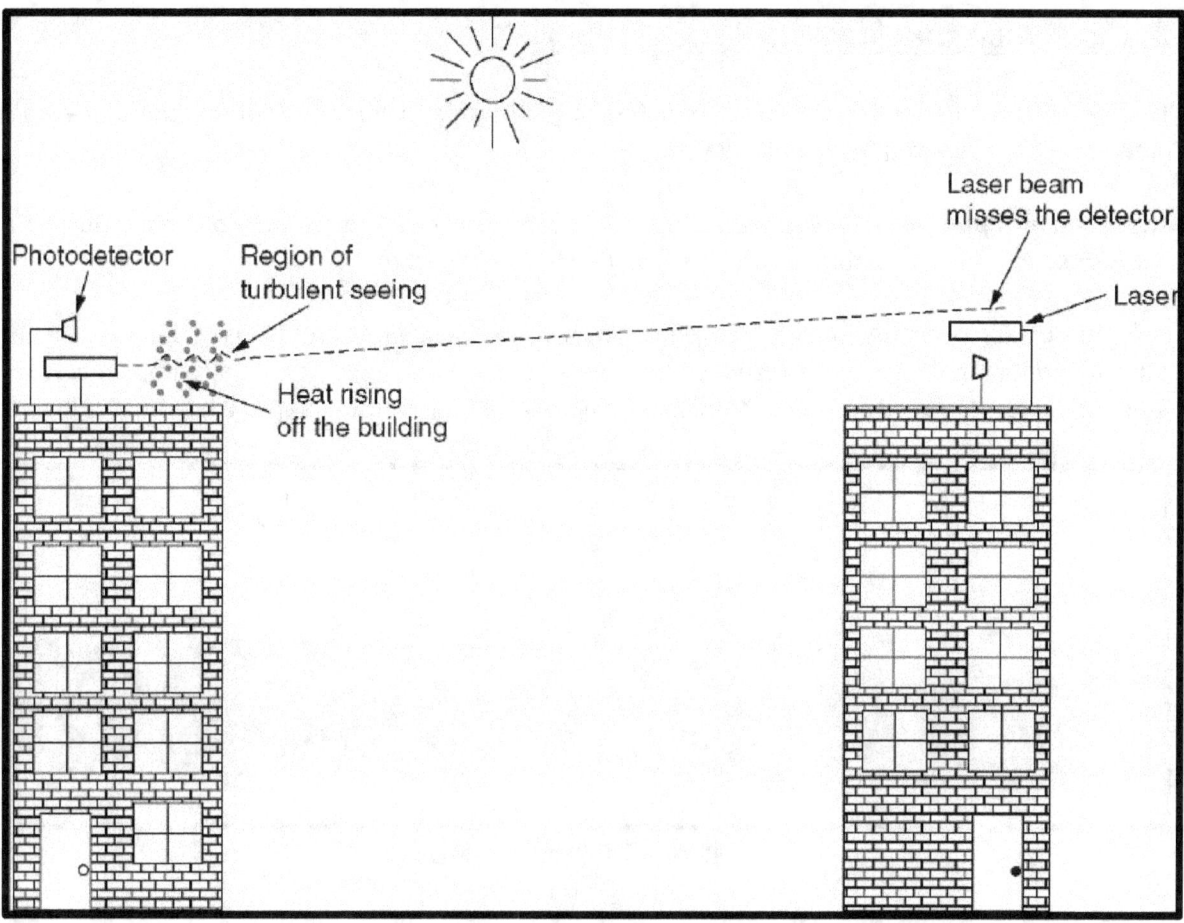

Figure 1.12: Light Wave Transmission

1.1.5 Multiplexing:

To make efficient use of the transmission system, it is desirable to **carry multiple signals on a single medium**. This is referred to as **multiplexing**.

There are **n inputs** to a **multiplexer**. The multiplexer is connected by a **single data link** to a **demultiplexer**. The link is able to carry **n separate channels of data**.

The **multiplexer combines** (multiplexes) data from the **n input lines** and transmits over a higher capacity data link. The **demultiplexer** accepts the multiplexed data stream, **separates** (demultiplexes) **the data** according to channel, and delivers them to the appropriate output lines.

Figure 1.13: Multiplexing

Two techniques for multiplexing in telecommunications networks are in common use: **frequency division multiplexing (FDM)** & **time division multiplexing (TDM)**.

FDM takes advantage of the fact that the useful bandwidth of the channel exceeds the required bandwidth of a given signal. A number of signals can be carried simultaneously if each signal is modulated onto a different carrier frequency and the carrier frequencies are sufficiently separated so that the bandwidths of the signals do not overlap.

TDM takes advantage of the fact that the achievable bit rate of the medium exceeds the required data rate of a digital signal. Multiple digital signals can be carried on a single frequency by dividing it into different time slots, interleaving portions of each signal in time.

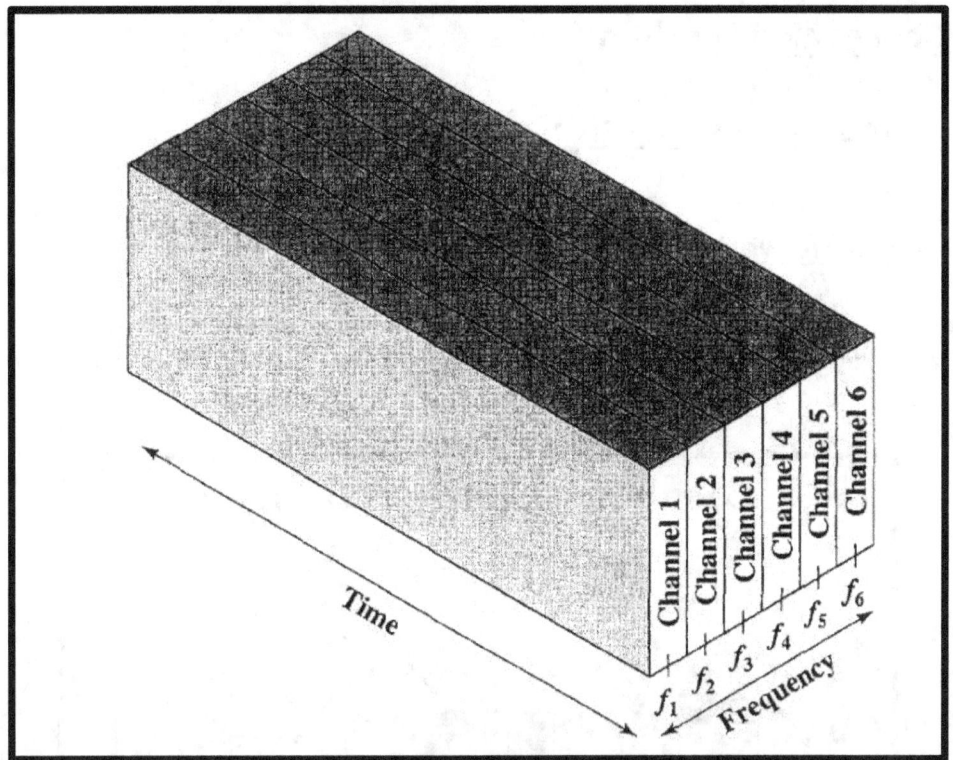

Figure 1.14: Frequency Division Multiplexing

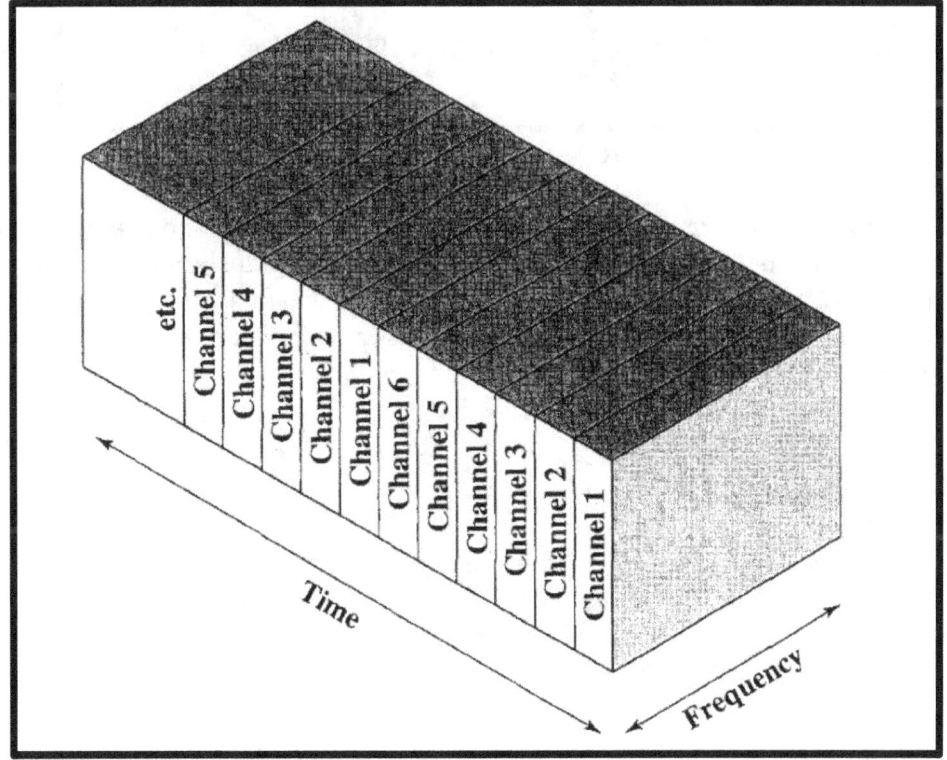

Figure 1.15: Time Division Multiplexing

1.2 Communication Networks:

1.2.1 LANs, MANs & WANs:

LAN (Local Area Network)

LAN is a **privately-owned** network within a **single building or campus** of up to a few kilometers in size. They are widely used to connect personal computers and workstations in company offices and factories to share resources (e.g., printers) and exchange information.

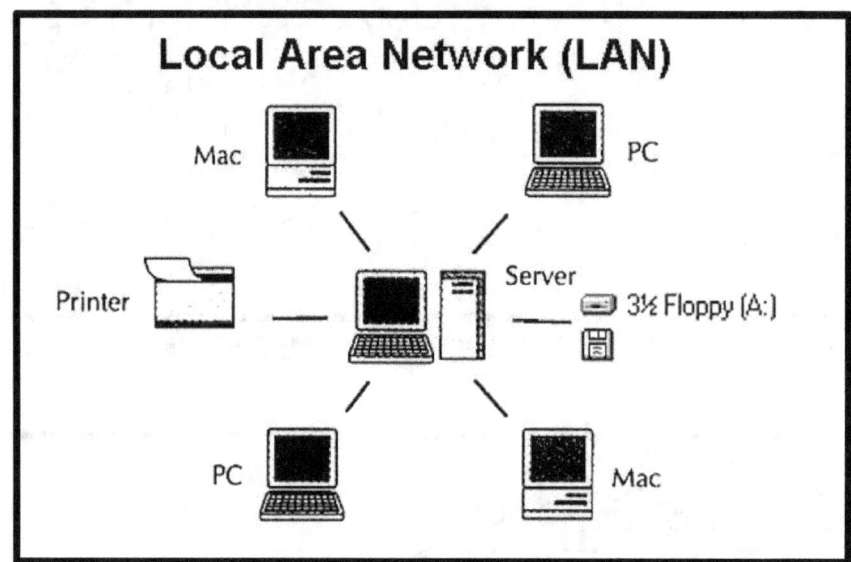

Figure 1.16: Local Area Network

LANs are **easy to design and troubleshoot**. In LAN, all the machines are connected to a **single cable**. Different types of **topologies** such as **Bus, Ring, Star, and Tree** are used. The **data rates for LAN** range from **4 to 16 Mbps**. They transfer data at high speeds (higher bandwidth) & exist in a **limited geographical area**. Connectivity, transmission media & resources are usually managed by the company which running the LAN.

MAN (Metropolitan Area Network)

A **Metropolitan Area Network**, or MAN, **covers a city**. The best-known example of a MAN is the **cable television network** available in many cities. A MAN is basically a **bigger version of a LAN** and normally uses similar technology. At first, the companies began jumping into the business, getting contracts from city governments to wire up an entire city. The next step was television programming and even entire channels designed for cable only. Often these channels were highly specialized, such as all news, all sports, all cooking, and so on.

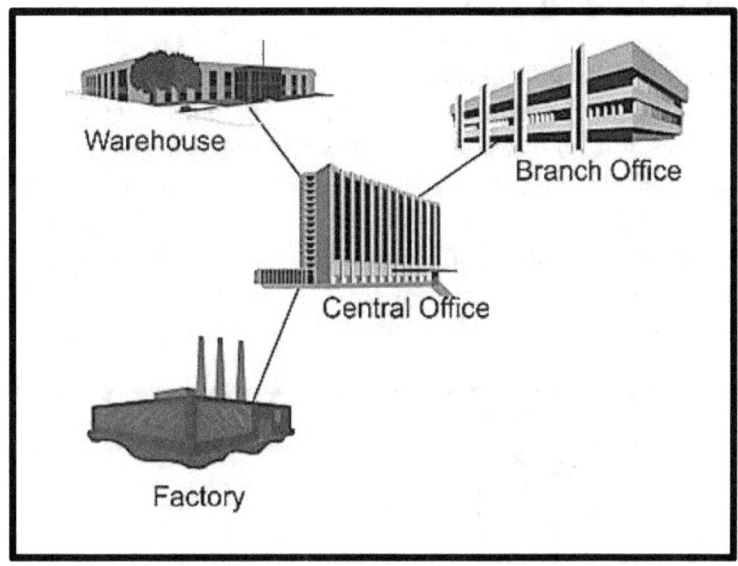

Figure 1.17: Metropolitan Area Network

WAN (Wide Area Network)

WAN, spans a large geographical area, often a **country or continent**. It contains a collection of machines intended for running user programs. We will follow traditional usage and call these machines as **hosts**. The **hosts** are **connected by a subnet**. In most wide area networks, the **subnet** consists of **two distinct components**: **transmission lines** and **switching elements**. Transmission lines move bits between machines. The communication between different users of WAN is established using leased telephone lines or satellite links and similar channels.

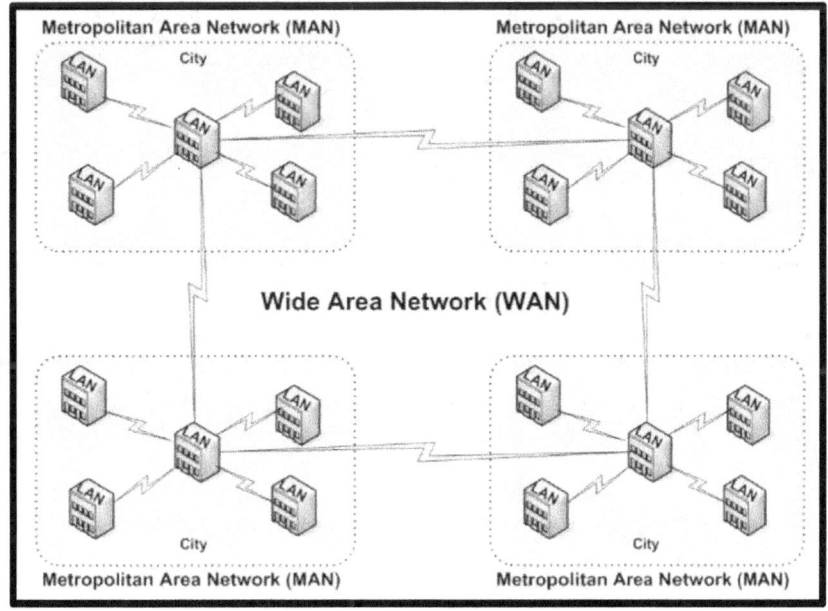

Figure 1.18: Wide Area Network

1.2.2 Switching Techniques:

For **transmission of data beyond a local area**, communication is typically achieved by transmitting data from source to destination through a **network of intermediate switching nodes**; this switched network design is sometimes used to **implement LANs and MANs** as well. The purpose of **switching nodes** is to provide a switching facility that will **move the data from node to node until they reach their destination**. The **end devices** that wish to communicate may be referred to as **stations**. We will refer to the **switching devices** whose purpose is to provide communication as **nodes**. The **nodes are connected** to each other in some topology by **transmission links**. The below figure demonstrates a **simple switching network**:

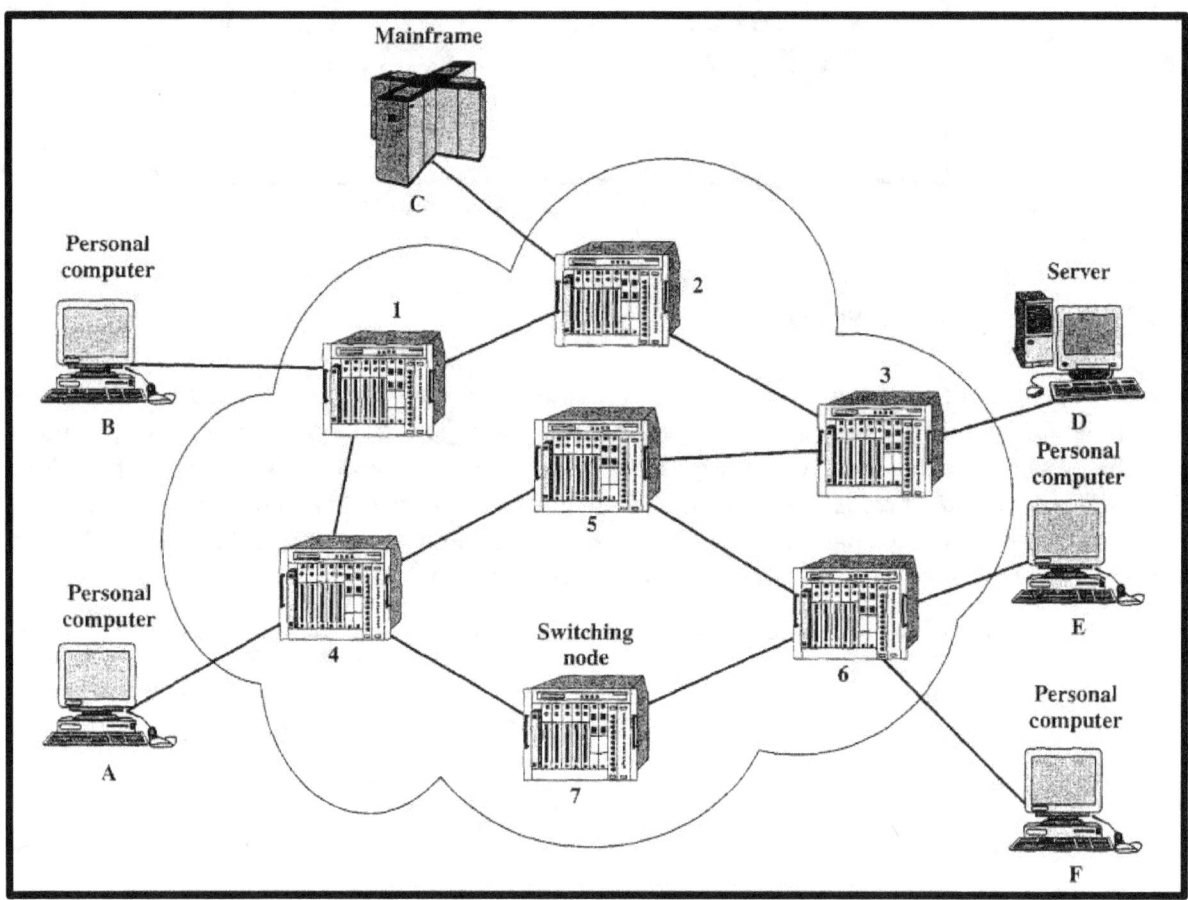

Figure 1.19: Simple Switching Network

Two quite different technologies are used in switching networks: **circuit switching** and **packet switching**. These two technologies differ in the way the nodes switch information from one link to another.

1.2.3 Circuit Switching:

Communication via circuit switching implies that there is a ***dedicated communication path*** between two stations. That path is a ***connected sequence of links*** between network nodes. On each physical link, a ***channel is dedicated*** to the connection. The most common ***example*** of circuit switching is the ***telephone network***. Circuit Switching network was ***specifically designed*** to handle ***voice traffic***.

Communication via circuit switching involves ***three phases*** that are explained below with reference to figure 19:

1. ***Circuit Establishment***: Before any signals can be transmitted, an end-to-end (station-to-station) circuit must be established. For example, station A sends a request to node 4 requesting a connection to station E. Typically, the link from A to 4 is a dedicated line, so that part of the connection already exists. Node 4 must find the next leg in a route leading to E. Based on routing information, node 4 selects the link to node 5, allocates a free channel on that link, and sends a message requesting connection to E. The remainder of the process proceeds similarly. Node 5 dedicates a channel to node 6 and internally ties that channel to the channel from node 4. Node 6 completes the connection to E.
2. ***Information Transfer***: Information can now be transmitted from A through the network to E. The path is A-4link, internal switching through 4, 4-5 channel, internal switching through 5, 5-6 channel, internal switching through 6, 6-E link. Generally, the connection is full duplex, and signals may be transmitted in both directions simultaneously.
3. ***Circuit Disconnect:*** After some period of information transfer, the connection is terminated, usually by the action of one of the two stations. Signals must be propagated to nodes 4,5, and 6 to deallocate the dedicated resources.

Advantages:
- Once the circuit is established, the network is effectively transparent to the users.
- Information is transmitted at a fixed data rate with no delay other than the propagation delay.

Disadvantages:
- Circuit switching can be inefficient because the channel capacity is dedicated for the entire duration of a connection. If no data is being transferred at a given time, the channel capacity gets wasted.
- Connection establishment between two stations takes time.

1.2.4 Packet Switching:

Packet switching was designed to provide a **more efficient** facility than circuit-switching for handling huge **data traffic**. With packet switching, a station transmits **data in small blocks, called packets**.

At each node packets are received, stored briefly (buffered) and passed on to the next node. This is known a **Store and forward mechanism.**

Each packet contains some portion of the **user data (payload) + control information (header)** needed for proper functioning of the network. A key element of **packet-switching networks** is whether the internal operation is **datagram** or **virtual circuit (VC)**.

With **internal VCs, a route is defined between two endpoints** and **all packets for that VC follow the same route.**

With **internal datagrams, each packet is treated independently**, and **packets intended for the same destination may follow different routes**.

Examples of packet switching networks are **X.25, Frame Relay, ATM** and **IP**.

Station breaks long message into packets. Packets are sent one at a time to the network. **Packets handled in two ways**:

1. **Datagram**:
    - Each packet treated independently.
    - Packets can take any practical route.
    - Packets may arrive out of order.
    - Packets may go missing.
    - It is up to the receiver to re-arrange packets in to correct order.

2. **Virtual Circuit**:
    - Preplanned route established before any packet is sent.
    - Once route is established, all the packets between the two communicating parties follow the same route through the network.
    - Call request packets and Call accept packets establish connection (handshake).
    - Each packet contains a Virtual Circuit Identifier (VCI) instead of destination address.
    - No routing decisions required for each packet.
    - The path is not a dedicated path.

Advantages of Packet Switching over Circuit Switching:

- Efficiency is greater, since a single node-to-node link can be dynamically shared by many packets over time.

- Priorities can be used. Thus, if a node has a number of packets queued for transmission, it can transmit the higher-priority packets first.
- When traffic becomes heavy on a circuit-switching network, some calls are blocked; that is, the network refuses to accept additional connection requests until the load on the network decreases. On a packet-switching network, packets are still accepted, but delivery delay increases.

Disadvantages of Packet Switching over Circuit Switching:

- To route packets through the network, overhead information such as address of the destination must be added to each packet, which reduces the space available for carrying user data.
- More processing is involved at each node in the transfer of information using packet switching.

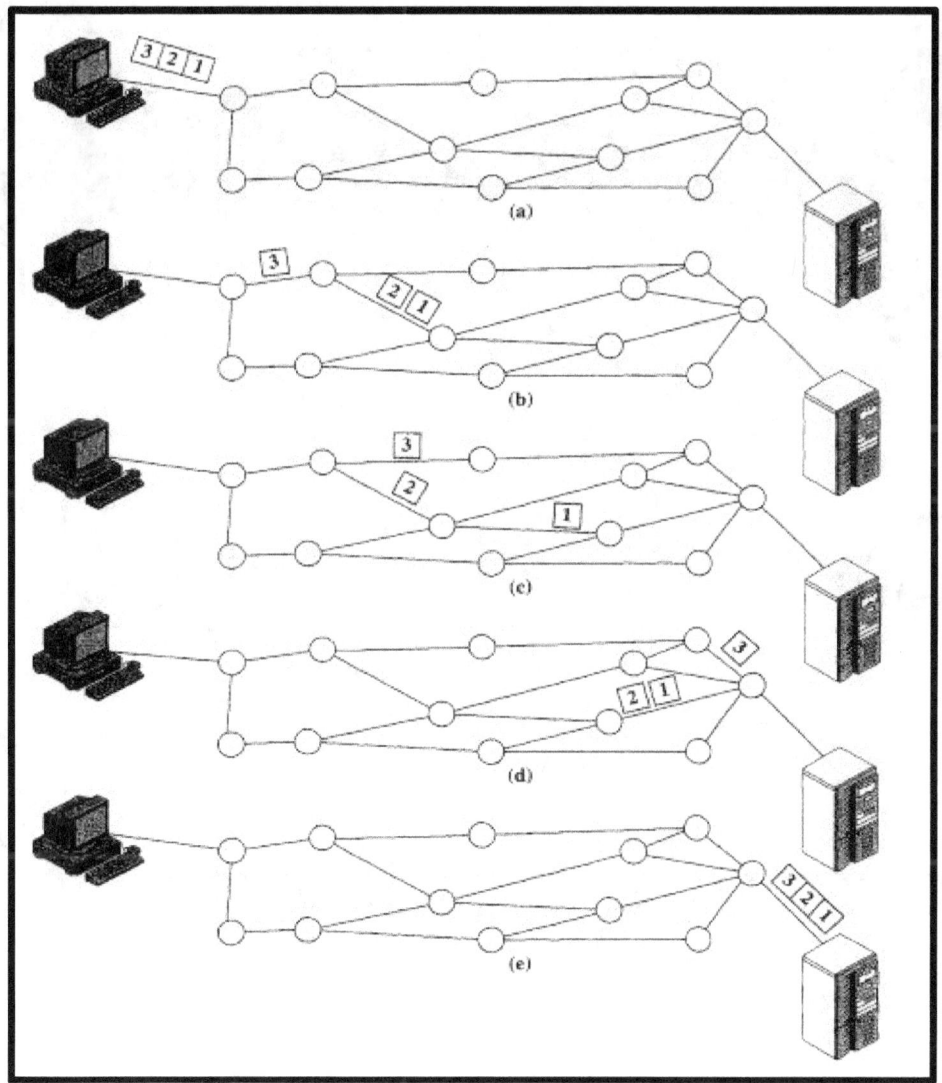

Figure 1.20: Packet Switching (Datagram)

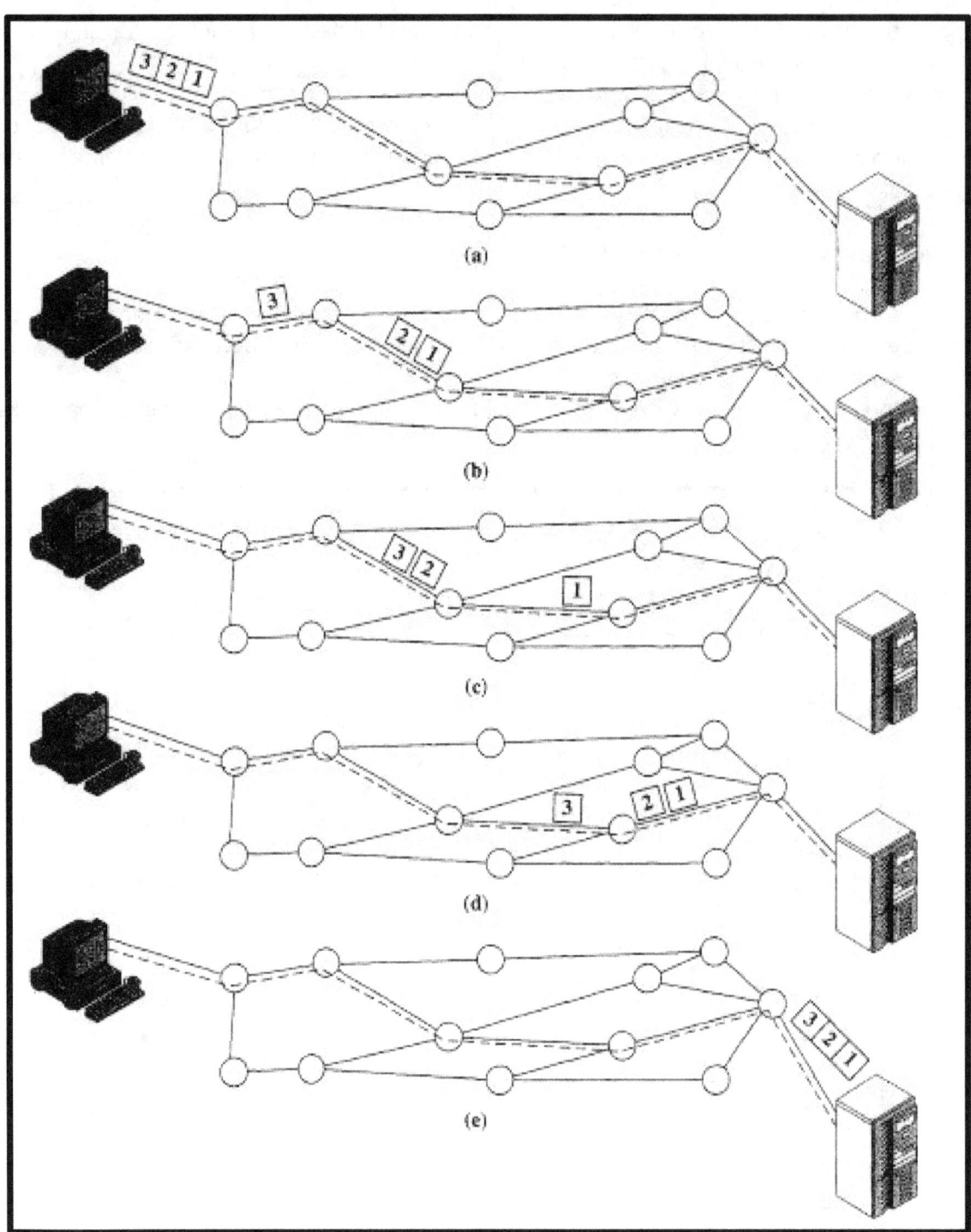

Figure 1.21: Packet Switching (Virtual Circuit)

1.3 Protocols and the TCP/IP Suite:

1.3.1 The Need for a Protocol Architecture:

When computers, terminals, and/or other data processing devices exchange data, the procedures involved can be quite complex. Consider, for example, the **transfer of a file between two computers**. **Typical tasks** to be performed include the following:

1. The source system must either activate the direct data communication path or inform the communication network of the desired destination system.
2. The source system must ascertain that the destination system is prepared to receive data.
3. The file transfer application on the source system must ascertain that the file management program on the destination system is prepared to accept and store the file for this particular user.
4. If the file formats used on the two systems are incompatible, one or the other system must perform a format translation function.

Instead of implementing the logic for these tasks as a single module, the task is broken up into subtasks, each of which is implemented separately. In a **protocol architecture**, the modules are arranged in a **vertical stack**. Each layer in the stack performs a subset of the tasks required for communication.

Ideally, layers should be defined so that **changes in one layer do not require changes in other layers**. Of course, it takes two to communicate, so the **same set of layered functions must exist in both the systems**.

The peer layers communicate by means of formatted blocks of data that obey a set of rules or conventions known as a **protocol**.

1.3.2 The TCP/IP Protocol Architecture:

Transmission Control Protocol/Internet Protocol (TCP/IP) protocol suite is primarily used for the Internet and networks worldwide. **TCP/IP** is a set of protocols developed to allow cooperating computers to share resources across the network. It has five layers:

- Application Layer
- Transport Layer
- Internet Layer
- Network Access Layer
- Physical Layer

1. **Application Layer**:

The **application layer** contains the logic needed to support the various user applications. For each different type of application, such as file transfer, a separate module is needed that is distinct to that application. The application layer in TCP/IP is equivalent to the session, presentation, and application layers combined in the OSI model.

2. **Transport Layer**:

Regardless of the nature of the applications that are exchanging data, there is usually a requirement that data be exchanged reliably. **Transport layer** assures that all of the data arrive at the destination application and in the same order in which they were sent. The most-used transport layer protocol is the **Transmission Control Protocol (TCP)**, which provides:

  - Connection-oriented reliable data delivery
  - Duplicate data suppression
  - Congestion control
  - Flow control.

Another **transport layer** protocol is the **User Datagram Protocol (UDP)**, which provides:
  - Unreliable data delivery.
  - Connectionless data transfer mechanism.
  - Faster data delivery

3. **Internet Layer**:

This layer is responsible for **internetworking**. In those cases where two devices are attached to different networks, procedures are needed to allow data to traverse through multiple interconnected networks. This is the function of the **internet layer**. The **Internet Protocol (IP)** is used at this layer to provide the routing function across multiple networks. This protocol is implemented not only in the end systems but also in routers.

NOTE: A **_router_** is a processor that connects two networks and whose primary function is to relay data from one network to the other on its route from the source to the destination end system.

4. **_Network Access Layer_**:

The **network access layer** is concerned with the **exchange of data** between an end system and the network to which it is attached. Basically, it is concerned with accessing and routing the data across a network for **two end systems which are attached to the same network**. This layer is the interface to the actual networking hardware. This interface may or may not provide reliable delivery, and may be packet oriented or stream oriented.

5. **_Physical Layer_**:

The **physical layer** covers the physical interface between a data transmission device and a transmission medium or network. This layer is concerned with specifying the characteristics of the transmission medium, the nature of the signals, the data rate, and related matters.

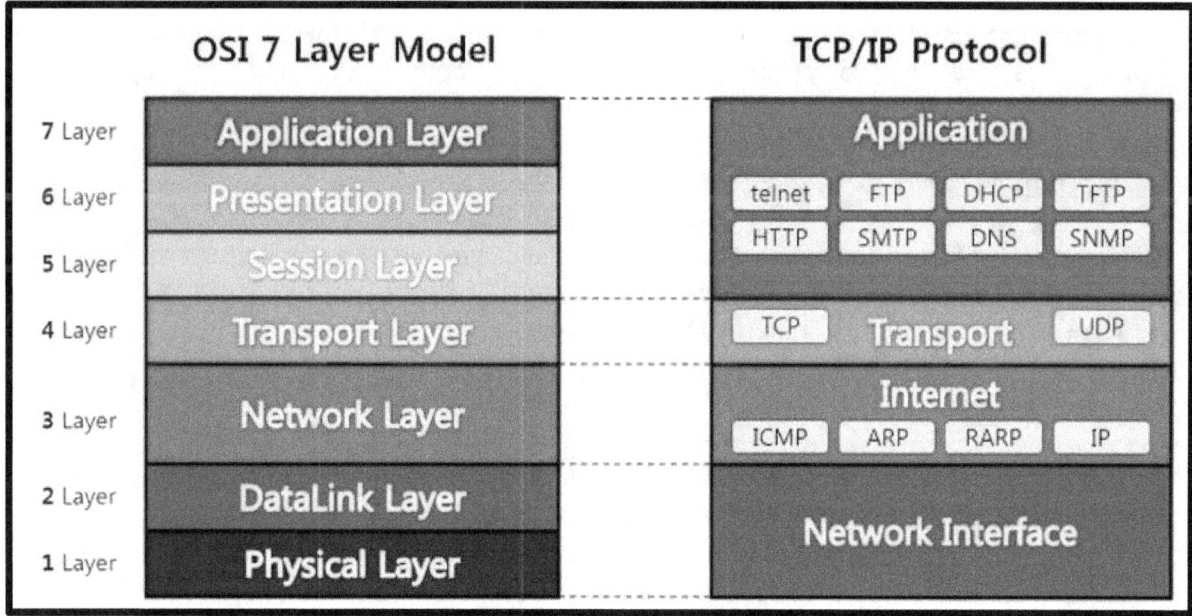

Figure 1.22: TCP/IP Architecture

1.3.3 The OSI Model:

OSI model is based on a proposal developed by the **International Standards Organization (ISO)** as a first step toward **international standardization of the protocols** used in the various layers.

The model is called the **ISO OSI (Open Systems Interconnection) Reference Model** because it deals with connecting open systems—that is, systems that are open for communication with other systems.

OSI model consists of **seven layers**:
- Application
- Presentation
- Session
- Transport
- Network
- Data link
- Physical

1. **Application Layer**:

The **application layer** enables the user to access the network. It provides user interfaces and support for services such as electronic mail, remote file access and transfer, shared database management, and other types of distributed information services.

2. **Presentation Layer**:

The **presentation layer** is concerned with the syntax (language rule) and semantics (meaning of each rule) of the information exchanged between two systems. The presentation layer is concerned with the Translation, Encryption & Compression of the data.

3. **Session Layer**:

The **session layer** is the **network dialog controller**. It **establishes, manages, and terminates connections** (sessions) between cooperating applications.

4. **Transport Layer**:

Transport layer provides reliable, transparent **transfer of data between end points.** The transport layer **ensures** that the whole **message arrives intact and in order**. It provides the following services:

- Segmentation and reassembly
- Connection control
- Flow control
- End-to-End error control

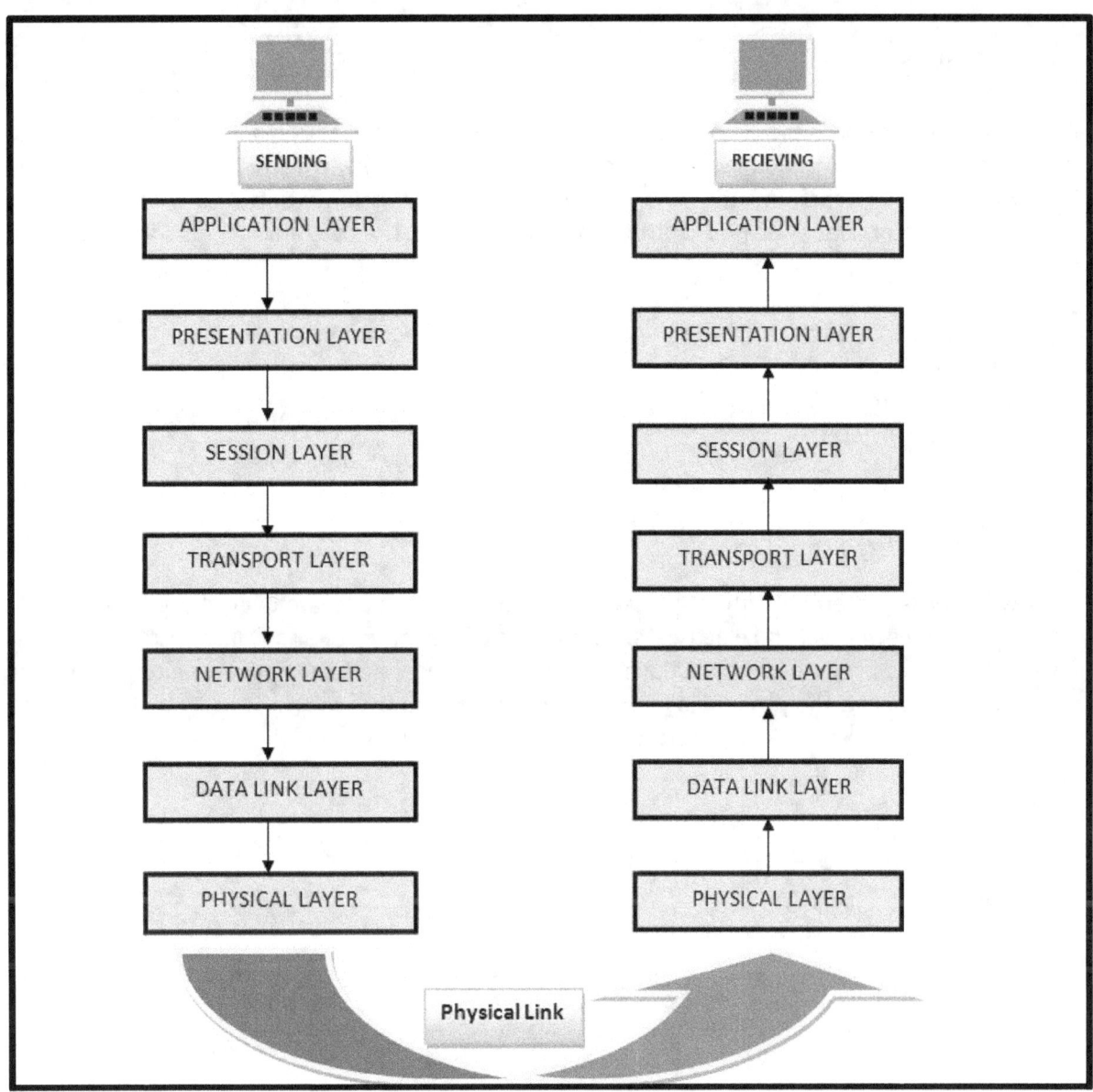

Figure 1.23: OSI Architecture

5. *Network Layer*:

The **network layer** is responsible for the **source-to-destination delivery** of a packet, possibly **across multiple networks**. The network layer is concerned with the **Logical addressing & Routing**. Network layer ensures that each data packet gets from its point of origin to its final destination.

If two systems are connected to the same link, there is usually no need for a network layer. However, if the **two systems are attached to different networks** (links) with connecting devices between the networks (links), there is often a need for the **network layer** to accomplish source-to-destination delivery.

6. ***Data Link Layer***:

Data link layer is responsible for providing reliable transfer of information across the physical link. It sends blocks (frames) of data with the necessary synchronization, error control, and flow control. The **data link layer** transforms the physical layer, a raw transmission facility, into a reliable link. It makes the physical layer appear error-free to the upper layer. The data link layer is concerned with the following:
- Framing
- Physical addressing
- Flow control
- Error control
- Access control

7. ***Physical Layer***:

Physical layer is concerned with transmission of unstructured bitstream over physical medium. It deals with the mechanical and electrical specifications of the interface and transmission medium. It also defines the procedures and functions that physical devices and interfaces have to perform for transmission to occur. The physical layer is concerned with the following:
- Physical characteristics of interface and medium
- Representation of bits
- Data rate
- Synchronization of bits
- Line configuration
- Physical topology
- Transmission mode

1.3.4 OSI versus TCP/IP:

OSI	TCP/IP
Stands for Open Systems Interconnection.	Stands for Transmission Control Protocol / Internet Protocol.
OSI is a generic, protocol independent standard.	TCP/IP model is based on standard protocols around which the Internet has developed.
OSI has 7 layers.	TCP/IP has 5 layers.
OSI has strict boundaries & it clearly defines its services, interfaces & protocols.	TCP/IP does not have very strict boundaries & it does not clearly define its services, interfaces & protocols.
In the network layer, OSI supports both connectionless and connection-oriented communication.	In the network layer, TCP/IP supports only connectionless communication.
In OSI model the transport layer guarantees the delivery of packets.	In TCP/IP model the transport layer does not guarantees delivery of packets.
OSI follows vertical approach.	TCP/IP follows horizontal approach.
OSI model has a separate Presentation layer and Session layer.	TCP/IP does not have a separate Presentation layer or Session layer.
OSI is a reference model around which the networks are built. It is generally used as a guidance tool.	TCP/IP model, in a way, is implementation of the OSI model.
Protocols are hidden in OSI model and are easily replaced as the technology changes.	In TCP/IP replacing protocol is not easy.

1.3.5 Internetworking:

In most cases, a LAN or WAN is not an isolated entity. An organization may have more than one type of LAN at a given site to satisfy a spectrum of needs.

- An organization may have multiple LANs of the same type at a given site to accommodate performance or security requirements.
- And an organization may have LANs at various sites and need them to be interconnected via WANs for central control of distributed information exchange.

An interconnected set of networks, from a user's point of view, may appear simply as a larger network. Following are some terminologies related to internetworking:

1. **_Communication Network_**: Facility that provides data transfer service among devices attached to the network.
2. **_Internet_**: Collection of communications networks interconnected by bridges and/or routers.
3. **_Intranet_**: Corporate internet operating within the organization.
4. **_End System (ES)_**: Device attached to one of the networks of an internet. Supports end-user applications or services.
5. **_Intermediate System (IS)_**: A device used to connect two networks & permit communication between End Systems attached to different networks. Intermediate Systems provide a communications path and perform the necessary relaying and routing functions so that data can be exchanged between devices attached to different subnetworks in the internet. Two types of ISs of particular interest are bridges and routers
6. **_Bridge_**: An intermediate system used to connect two LANs using similar LAN protocols. It acts as an address filter, picking up packets from one LAN that are intended for destination on another LAN & passing on those packets. It operates on layer 2 (Data Link Layer) of the OSI model.
7. **_Router_**: An intermediate system used to connect two networks that may or may not be similar. The router employs an internet protocol present in each router & each end system of the network. It operates on layer 3 (Network Layer) of the OSI model.

Chapter 2

Introduction to Mobile Computing

2.1 Mobile Computing:

2.1.1 Introduction to Mobile Computing & it's Principles:

Mobile Computing is a technology that allows transmission of data, voice and video via a wireless device without having to be connected to a fixed physical link.

Following are the *principles of mobile computing*:

- **Portability**: Facilitates movement of device(s) within the mobile computing environment. These devices may have limited device capabilities and limited power supply, but should have a sufficient processing capability and physical portability to operate in a movable environment.
- **Connectivity**: This defines the quality of service (QoS) of the network connectivity. In a mobile computing system, the network availability is expected to be maintained at a high level with the minimal amount of lag/downtime without being affected by the movement of the connected nodes.
- **Social Interactivity**: Maintaining the connectivity to collaborate with other users (nodes), at least within the same environment.
- **Individuality**: Adapting the technology to suit individual needs. A portable device or a mobile node connected to a mobile network often denote an individual; a mobile computing system should be able to adopt the technology to cater the individual needs and also to obtain contextual information of each node.

Wireless devices that are usually used for mobile computing are: PDAs (Personal Digital Assistants), Portable Computers, Tablet Computers, Smart Phones, and so on.

2.1.2 Architecture of Mobile Computing:

The *three-tier architecture of mobile computing* contains the *user interface or the presentation tier*, *the process management or the application tier* and *the data management tier*.

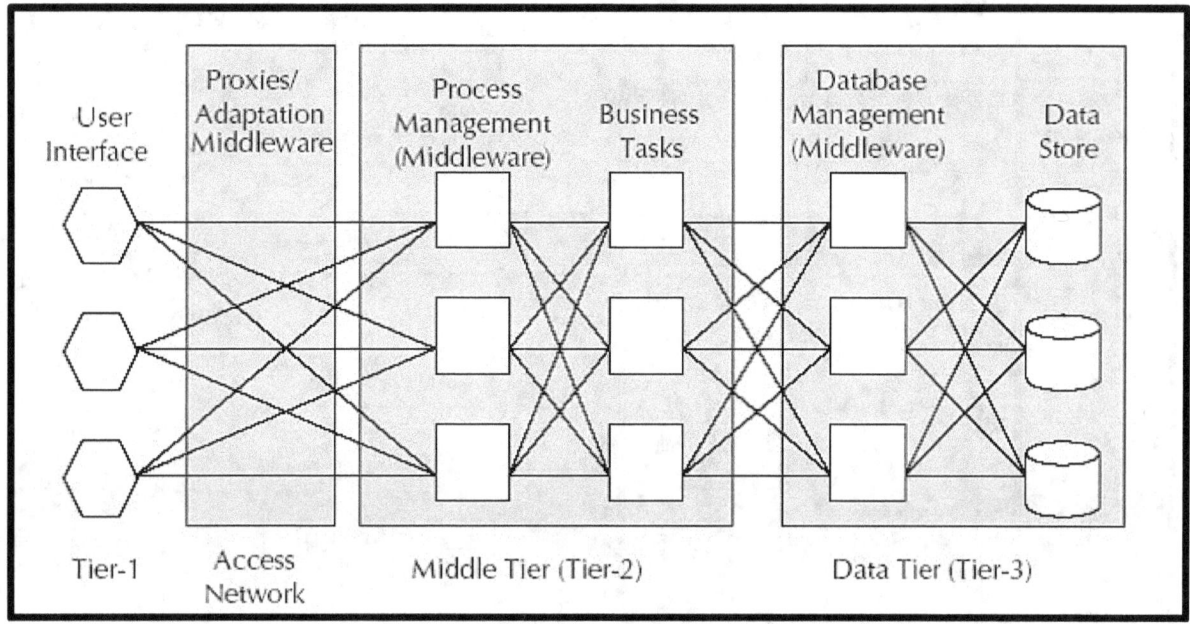

Figure 2.1: Architecture of Mobile Computing

A. **Presentation Tier:**

- This layer deals with user facing device handling and rendering.

- This tier includes a user system interface where user applications resides. These applications run on the client device and offer all the user interface.

- Presentation Tier is responsible for presenting the information to the end user.

- It includes web browsers (like Mozilla, Internet Explorer, etc.), WAP browsers and customized client programs.

B. **Application Tier:**

- This layer is capable of accommodating hundreds of users.

- This tier implements the business logic of processing user input, obtaining data, and making decision

- The process management middleware in this tier controls transaction and asynchronous queuing to ensure reliable completion of transaction.

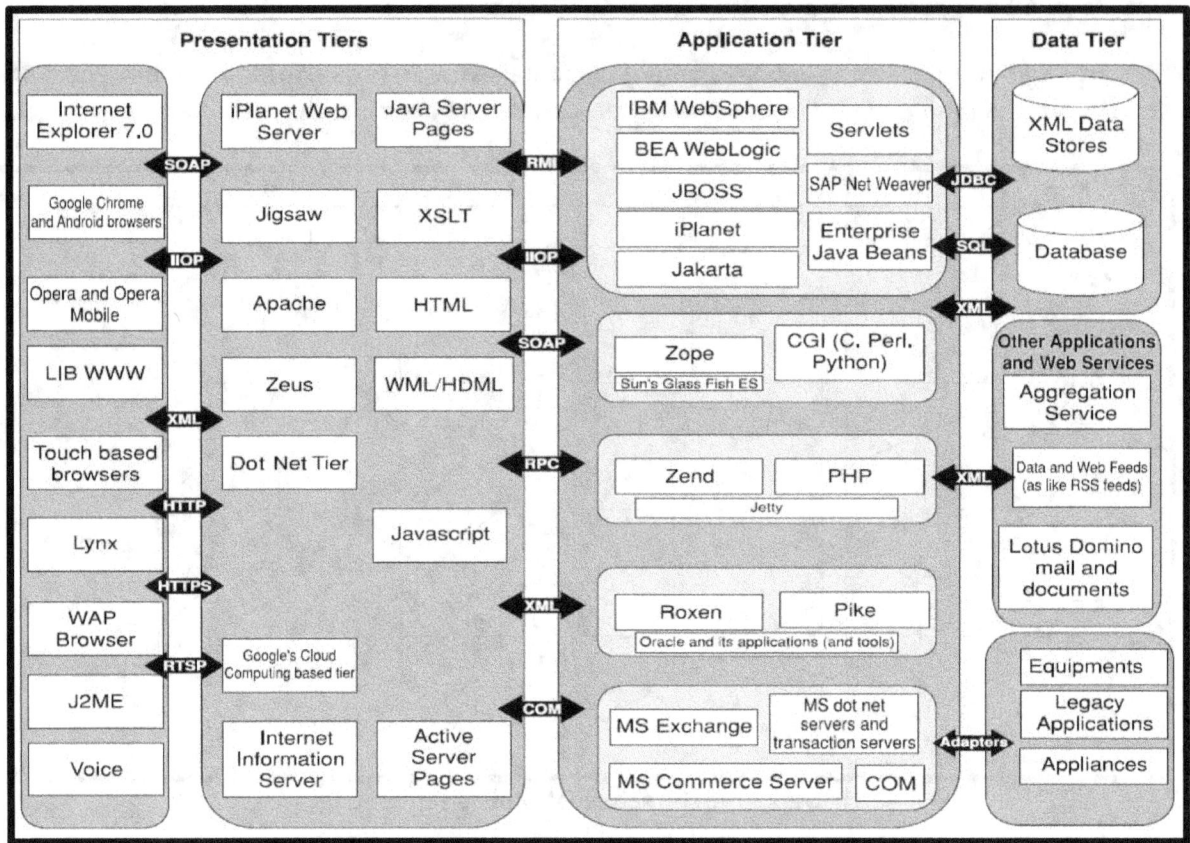

Figure 2.2: 3-Tier Architecture of Mobile Computing

C. **<u>Data Management Tier</u>**:

- o The Data Management Tier is used to store data needed by the application and acts as a repository for both temporary and permanent data.

- o These data can range from sophisticated relational databases to simple text files.

- o Database middleware allows the business logic to be independent and transparent of the database technology and the database vendor. It runs between the application program and the database. Using this middleware, the application will be able to access data from any source.

- o Examples of database middleware are ODBC & JDBC.

The three-tier architecture is better suited for an effective networked client/server design. It provides increased performance, flexibility, maintainability, reusability and scalability, while hiding the complexity of distributed processing from user. All these characteristics have made three-tier architecture a popular choice for Inter application and net-centric information system.

2.1.3 Advantages of Mobile Computing:

Following are the *major advantages of Mobile Computing*:

- **A. *Location Flexibility:*** This has enabled users to work from anywhere as long as there is a connection established. A user can work without being in a fixed position. Their mobility ensures that they are able to carry out numerous tasks at the same time and perform their stated jobs.

- **B. *Saves Time:*** The time consumed or wasted while travelling from different locations or to the office and back, has been slashed. One can now access all the important documents and files over a secure channel or portal and work as if they were on their computer. It has also reduced unnecessary incurred expenses.

- **C. *Enhanced Productivity:*** Users can work efficiently and effectively from whichever location they find comfortable. This in turn enhances their productivity level.

- **D. *Ease of Research***: Research has been made easier, since users earlier were required to go to the field and search for facts and feed them back into the system. Mobile computing has made it easier for field officers and researchers to collect and feed data from wherever they are without making unnecessary trips to and from the office to the field.

- **E. *Entertainment***: Video and audio recordings can now be streamed on-the-go using mobile computing. It's easy to access a wide variety of movies, educational and informative material. With the improvement and availability of high speed data connections at considerable cost, one is able to get all the entertainment they want as they browse the internet for streamed data. One is able to watch news, movies, and documentaries among other entertainment over the internet. This was not possible before mobile computing dawned on the computing world.

- **F. *Streamlining of Business Processes***: Business processes are now easily available through secured connections. Meetings, seminars and other informative services can be conducted using video and voice conferencing. Travel time and expenditure is also considerably reduced.

2.1.4 Security Issues in Mobile Computing:

Security Issues	Description
Confidentiality	Only the intended user must be allowed to read data. It should be hidden from all other parties. (Encryption is a method used to solve it)
Integrity	Data integrity is concerned with the correctness of data. The data needs to have integrity or else user receives a manipulated message.
Spoofing	A node can impersonate (pretend to be) an address in a mobile ad hoc network
Availability	Attacks similar to denial of service can block the availability of data at the user end. (E.g. an intermediate router can be configured to attack packets and stop them or re-route them.)
Pre-keying	In case of encrypted system, key exchange is necessary before the actual data transfer. Now if this key (especially if private-key) is send via a wireless network, there can be an issue of key-trapping
Resource constraints	An attack may sometimes cripple the resources available to a mobile system like limited battery, slower CPUs, exhausting memory due to caching, etc. Such resources may get drained or may sometimes be not be strong enough to fight back.
Intercepted	Wireless signals, since being transferred via open-air can be intercepted.
Replaying	After carefully analyzing the authentication requests and client responses, an attacker can replay such a similar sequence again.
Stealing	If an attacker steals the user-id and password of a subscriber, or gets his SIM card, he/she will be able to enjoy the user's subscriptions.
Mobility concerns	When a MS (Mobile Station) moves from one cell to another, the connection will be routed though different paths which cannot be relied upon.
Eavesdropping	Peeping into someone else's conversation is called eavesdropping. (e.g. phone tapping)

The various solutions that exist to solve the various problems affecting mobile computing are:

Direct Signaling	**We can use directed signals which are just sufficient to reach the user's device and establish a proper-link with him. This helps prevent security risks coming from other directions and also at farther distances in same line.**
Hashing	Hashing is a method employed to check the data integrity. A hash function is applied on the actual data resulting in some bits of data (integer value). In case of a manipulation by a third-party, the hash value will be altered.
MAC	Message Authentication Code is a combination of hash and secret key. For extra added security
Encryption	Encryption is conversion of code into a cipher text understood only by a person having the decryption key. It includes both the public key and private key method. Some examples are DES, AES, Ceaser-cipher, RSA etc.
Checksum or Parity	These are the basic methods used for checking data integrity (counting the total number of 1's or 0's)
SSL	A very famous feature used in today's communication systems is SSL. SSL stands for secure socket layer. It is a protocol that runs between HTTP and TCP for secure transaction between client and Web server. Links using SSL protocols starts with HTTPS (https://www...). SSL supports hash function MD5 and SHA, digital signatures, RSA, various encryption algorithms.
IPsec	IPsec (internet protocol for security) contains various features for providing enhanced security. It includes an Authentication Header (AH- Packet Header focusing on Security), Encapsulating Security Payload (for confidentiality purposes) and Internet Key Exchange (IKE) (for secure exchange of keys used for encryption)
RADIUS	Remote Authentication Dial in User Service (RADIUS) is a networking protocol that provides centralized authentication, authorization, and Accounting (AAA) management for devices to connect and use a network service.

2.2 Cellular Wireless Networks:

2.2.1 Introduction to Cellular Networks:

In a **Cellular network**, rural and urban regions are subdivided into smaller areas called **"cells"**.
- **Each cell** can cover a limited number of **mobile subscribers** within its boundaries.
- **Each cell** can have **a base station** with a number of **RF (Radio Frequency) channels**.
- **Base station** is a **fixed-location transceiver** located within a cell. It contains a **radio transceiver and controller** & **provides radio communication to mobile units** located in cell.
- **Frequencies used in a given cell** area will be simultaneously **reused at a different cell** which is geographically separated.

For example, a typical seven-cell pattern can be considered as follow:

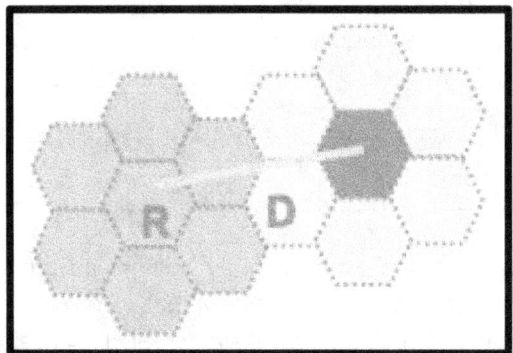

Figure 2.3: Cellular Network - 7 Cell Pattern

Total available frequency resources are **divided into seven parts** (each part consisting of a number of radio channels) and **allocated to a cell**. In a group of 7 cells, available frequency spectrum is consumed totally. The **same seven sets of frequencies can be used after certain distance**.

The group of cells where the available frequency spectrum is totally consumed is called a **cluster of cells. Frequency can't be reused within a cluster**.

Two cells having the same number in the adjacent cluster, use the same set of RF (Radio Frequency) channels and hence are termed as **"Co-channel cells"**. The distance between the cells using the same frequency should be sufficient to keep the **co-channel interference** to an acceptable level.

A cell might also contain **Mobile telephone switching offices (MTSO)**. The **MTSO links calls together** using traditional copper, fiber optic, or microwave technology. It also allows mobile communication devices in the cell to dial out and alerts devices in the cell of incoming calls. The **MTSO monitors the quality of the communications** signal and **transfers the call to another base station** which is better suited to provide communication to the mobile device.

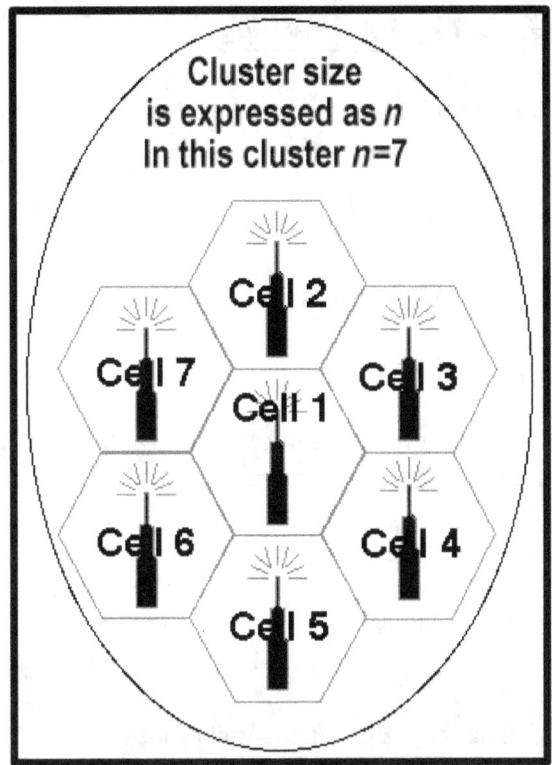

Figure 2.4: Cellular Network - Cluster

Shape of Cells:

For analytical purposes a "**Hexagon**" cell is preferred to other shapes on paper due to the following reasons.

- A *hexagon* layout *requires fewer cells* to cover a given area. Hence, it needs *fewer base stations* and *minimum capital investment*.

- Other geometrical shapes cannot effectively do this. For example, if *circular shaped* cells are there, then there will be *overlapping of cells*.

- Also for a given area, among square, triangle and hexagon, *radius of a hexagon will be the maximum* which is needed for weaker mobiles.

In reality, cells are not hexagonal but irregular in shape, determined by factors like propagation of radio waves over the terrain, obstacles, and other geographical constraints.

2.2.2 Principles of Cellular Networks:

Cellular network is an underlying technology for mobile phones, personal communication systems, wireless networking etc.

Following are the *principles/features of cellular networks*:

- Offer very high capacity in a limited spectrum.

- Reuse of radio channel in different cells.

- Enable a fixed number of channels to serve an arbitrarily large number of users by reusing the channel throughout the coverage region.

- Communication is always between mobile and base station (not directly between mobiles)

- Each cellular base station is allocated a group of radio channels within a small geographic area called a cell.

- Neighboring cells are assigned different channel groups.

- By limiting the coverage area to within the boundary of the cell, the channel groups may be reused to cover different cells.

- Keep interference levels within tolerable limits.

2.2.3 Frequency Reuse in Cellular Networks:

Because only a small number of radio channel frequencies were available for mobile systems, engineers had to find a way to reuse radio channels to *carry more than one conversation at a time*. The solution that was adopted was called *frequency planning* or *frequency reuse*.

It is the concept of using the same radio frequencies within a given area, that are separated by considerable distance, with minimal interference. That is, frequencies used in a given cell area will be simultaneously reused at a different cell which is geographically separated. The group of cells where the available frequency spectrum is totally consumed is called a *cluster of cells. Frequency can't be reused within a cluster*.

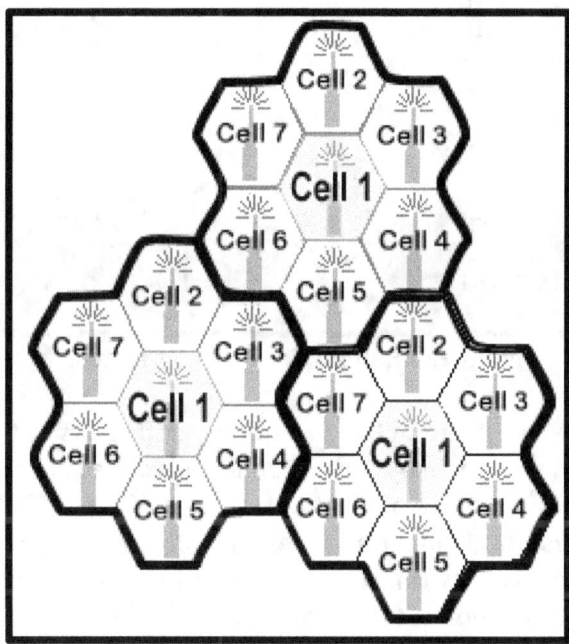

Figure 2.5: Frequency Reuse in Cellular Network

For example, consider the above figure. Cells with the *same number* have the *same set of frequencies*. Here, because the number of available frequencies is 7, the *frequency reuse factor* is 1/7. That is, each cell is using 1/7 of available cellular channels.

2.2.4 Handoff/Handover in Cellular Networks:

The final obstacle in the development of the cellular network involved the problem created when a mobile subscriber traveled from one cell to another during a call. As adjacent areas do not use the same radio channels, a call must either be dropped or transferred from one radio channel to another when a user crosses the line between adjacent cells. Because dropping the call is unacceptable, the process of handoff was created.

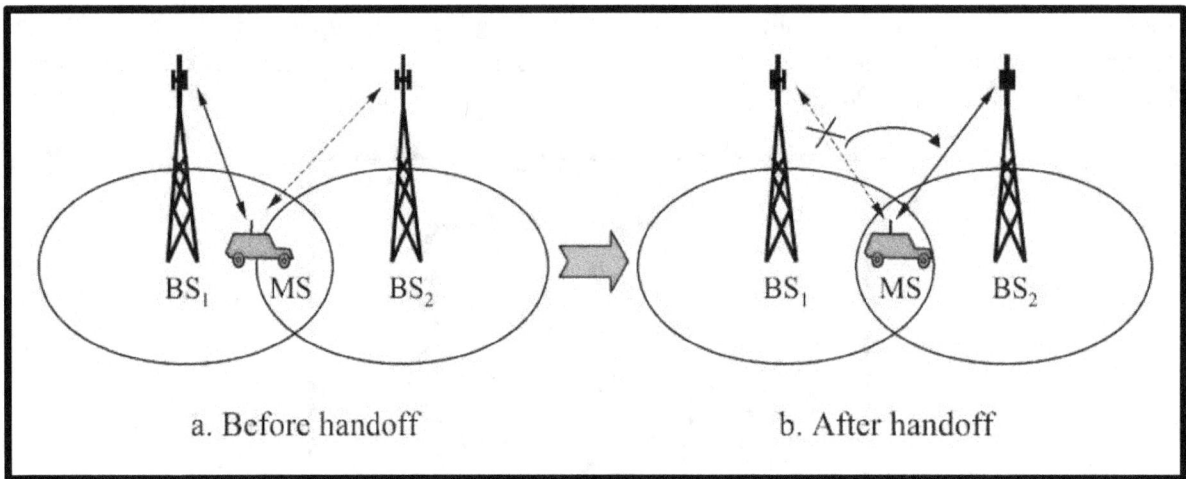

Figure 2.6: Handoff in a Cellular Network

In cellular telecommunications, the term handover or handoff refers to the process of transferring an ongoing call or data session from one channel connected to the core network to another. During a call, when the mobile unit moves out of the coverage area of a given cell, the reception becomes weak. At this point, the base station in use requests a handoff. The system switches the call to a stronger-frequency channel in a new cell without interrupting the call or alerting the user. The call continues and the user does not notice the handoff at all.

2.2.5 1G: First Generation Networks:

The **first mobile phone system** in the market was **AMPS (Advanced Mobile Phone System)**. It was the **first U.S. cellular telephone system**, deployed in Chicago in **1983**. The **main technology** of this first generation mobile system was **FDMA/FDD (Frequency Division Multiple Access/ Frequency Division Duplex)** & **Analog FM (Frequency Modulation)**. A total of **1664 channels** were available in the **824 to 849 MHz** & **869 to 894 MHz** frequency band, providing **832 downlink (DL)** & **832 uplink (UL) channels**.

AMPS supports frequency reuse. The underlying network is a **cellular network** where a geographic **region is divided into cells**. A **base station (BS) at the center** of the cell transmits signals to and from users within the cell.

2.2.6 2G: Second Generation Networks:

Digital modulation formats were introduced in this generation with the **main technology** as **TDMA/FDD (Time-Division Multiple Access/Frequency Division Duplex)** & **CDMA/FDD (Code-Division Multiple Access/Frequency Division Duplex)**. **Digital systems** make possible an array of new **services such as caller ID**.

The **Global System for Mobile Communications (GSM)** is a popular **2G system**. **GSM** offers a **data rate of 9.6 to 14.4 kbps**. It **supports international roaming**, which means users may have access to wireless services even when traveling abroad. The most **popular service offered by GSM** is the **Short Message Service (SMS)**, which allows users to send **text messages up to 160 characters long**.

2.2.7 2.5G Mobile Networks:

2.5G systems support more than just voice communications. In addition to text messaging, 2.5G systems offer a ***data rate on the order of 100 kbps*** to ***support various data technologies***, such as ***Internet access***. Most ***2.5G systems*** implement ***packet switching***. The 2.5G systems help provide seamless ***transition technology*** between ***2G and 3G*** systems. The following are ***2.5G systems***:

- ***High-Speed Circuit-Switched Data (HSCSD)***: Even though most 2.5G systems implement packet switching, HSCSD continues support for ***circuit-switched data***. It offers a ***data rate of 115 kbps*** and is designed to ***enhance GSM networks***. The ***access technology*** used is ***time division multiple access (TDMA)***. It provides ***support for Web browsing & file transfers***.

- ***General Packet Radio Service (GPRS)***: GPRS offers a ***data rate of 168 kbps***. It ***enhances the performance and transmission speeds of GSM***. ***GPRS*** provides ***always-on connectivity***, which means users do not have to reconnect to the network for each transmission. Because there is a ***maximum of eight slots to transmit calls on one device***, it allows more than one transmission at one time; for example, a voice call and an incoming text message can be handled simultaneously.

- ***Enhanced Data Rates for GSM Evolution (EDGE)***: ***EDGE*** works in ***conjunction with GPRS and TDMA over GSM networks***. Its offered ***data rate is 384 kbps***. EDGE supports data communications while voice communications are supported using the technology on existing networks.

2.2.8 3G: Third Generation Networks:

Third-generation (3G) wireless systems are designed to ***support high bit rate telecommunications***. 3G systems are designed to meet the requirements of ***multimedia applications*** and ***Internet services***. The ***bit rate offered ranges as follows:***

- *144 kbps for full mobility applications*
- *384 kbps for limited mobility applications in macro- and microcellular environments*
- *2 Mbps for low-mobility applications in micro- and pico-cellular environments*

A very useful service provided by 3G systems is an ***emergency service*** with the ability to ***identify a user's location***.

3G is based on the ***International Telecommunication Union (ITU)*** family of standards under the ***International Mobile Telecommunications-2000 (IMT-2000)***. Initially, the ***ITU*** intended to design a single 3G standard; however, due to a number of difficulties, it has ratified ***two 3G standards***. The two standards are:

1. ***CDMA 2000***, which provides a bit rate of up to 2.4 Mbps &
2. ***Wideband CDMA (WCDMA)***, which provides a bit rate of up to 8 Mbps.

The high bit rate enables new wireless services that can be classified into three categories:

- ***Information Retrieval***: It permits location-aware applications to remotely download information from a corporate database.
- ***Mobile Commerce***: It allows users to book a flight or pay bills.
- ***General Communication***: It permits users to make or receive phone calls, send or receive messages, or activate an air conditioner at home.

2.2.9 1G, 2G, 2.5G, 3G - Comparison:

The following diagram shows the evolution of cellular wireless networks:

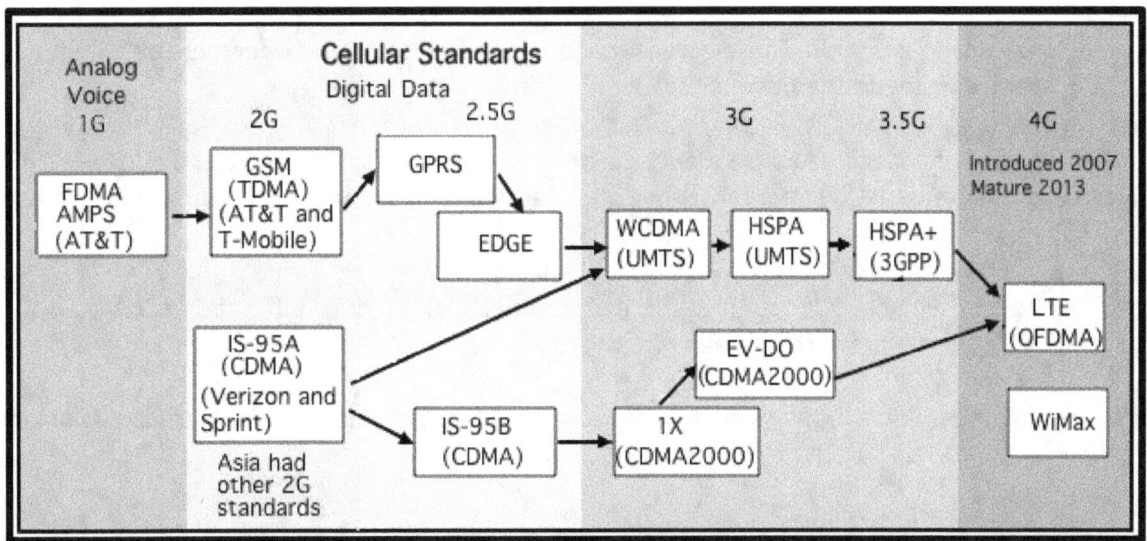

Figure 2.7: Evolution of Cellular Wireless Networks

1G	2G	2.5G	3G
The main technology in 1G was FDMA (Frequency Division Multiple Access) & Analog FM (Frequency Modulation).	The main technology in 2G was TDMA (Time Division Multiple Access) & CDMA (Code Division Multiple Access).	It's a transition technology between 2G and 3G systems. Most 2.5G systems implement packet switching	The main technology in 3G was CDMA (Code Division Multiple Access) & WCDMA (Wideband CDMA).
Data Rate: Less Than 10 Kbps	**Data Rate**: 9.6 Kbps to 64 Kbps	**Data Rate**: 64 Kbps to 144 Kbps	**Data Rate**: 384 Kbps to 2 Mbps
Example: AMPS	**Example**: GSM	**Example**: GPRS & EDGE	**Example**: ECDMA
Services: Poor Voice Quality, Poor Battery Life, Large Phone Size, No Security, Frequent Call Drops, Poor handoffs	**Services** – SMS (Short Message Service), Mobile Internet	**Services**: Data Communications & Enhanced Internet Services	**Services**: High Speed Internet services for Mobility, Global Roaming, Emergency Service (Identify user location), Video Conferencing, TV Streaming, Multimedia (Voice, Video, Data)

2.3 Antennas and Propagation:

2.3.1 Antennas:

An **antenna** can be defined as an electrical conductor or system of conductors used either for radiating electromagnetic energy or for collecting electromagnetic energy.

For transmission of a signal, radio-frequency electrical energy from the transmitter is converted into electromagnetic energy by the antenna and radiated into the surrounding environment (atmosphere, space, water).

For reception of a signal, electromagnetic energy impinging on the antenna is converted into radiofrequency electrical energy and fed into the receiver.

In two-way communication, the same antenna can be and often is used for both transmission and reception.

Types of Antenna:

- **Isotropic Antenna**: It is a point in space that radiates power equally in all directions. The actual radiation pattern for the isotropic antenna is a sphere with the antenna at the center.

- **Dipole Antennas:**

    - **Half-Wave Dipole Antenna**: It consists of two straight collinear conductors of equal length, separated by a small gap. The length of the antenna is one-half the wavelength of the signal that can be transmitted most efficiently.

    - **Quarter-Wave Vertical Antenna (or Marconi Antenna)**: It is the type of antenna commonly used for automobile radios and portable radios.

- **Parabolic Reflective Antenna**: It is used in terrestrial microwave and satellite applications. It is in the shape of a **parabola** & there is a fixed point called **focus**. If a parabola is revolved about its axis, the surface generated is called a **Paraboloid**. If a source of electromagnetic energy (or sound) is placed at the **focus** of the **paraboloid**, and if the **paraboloid** is a **reflecting surface**, then the **wave will bounce back in lines parallel to the axis** of the paraboloid; this effect creates a **parallel beam without dispersion**. The converse is also true. If incoming waves are parallel to the axis of the reflecting paraboloid, the resulting signal will be concentrated at the focus.

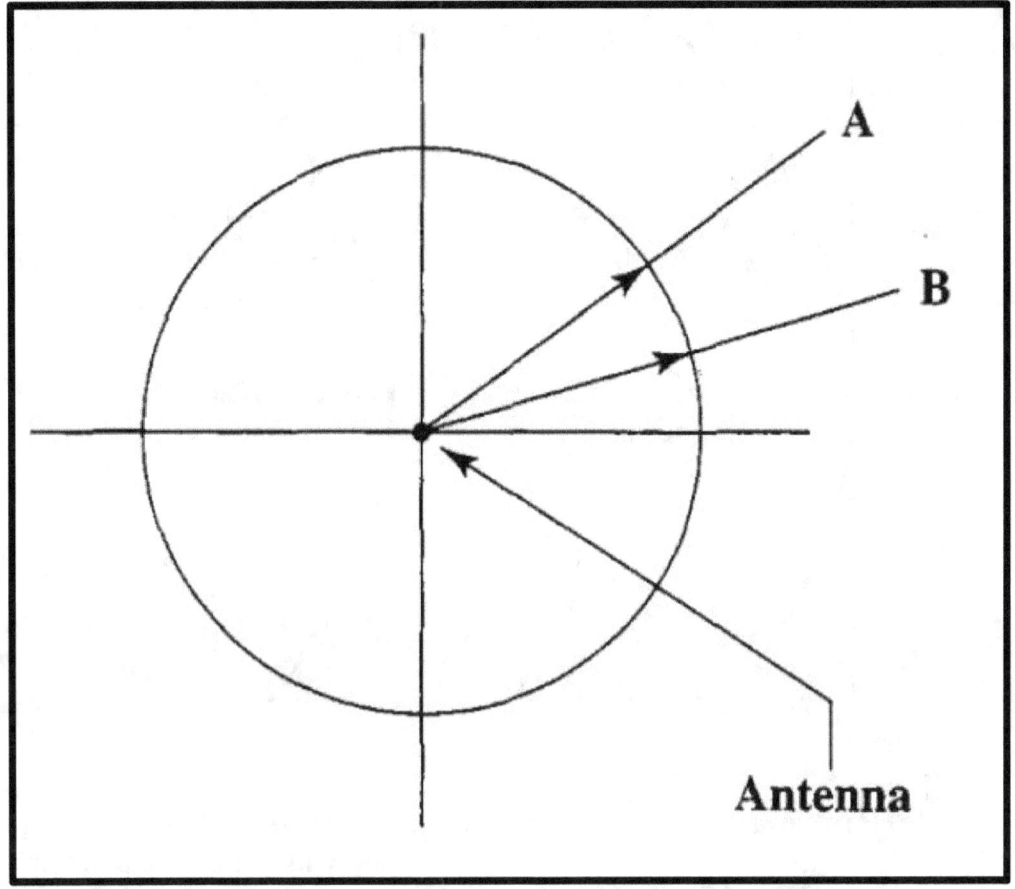

Figure 2.8: Radiation Pattern of an Isotropic Antenna

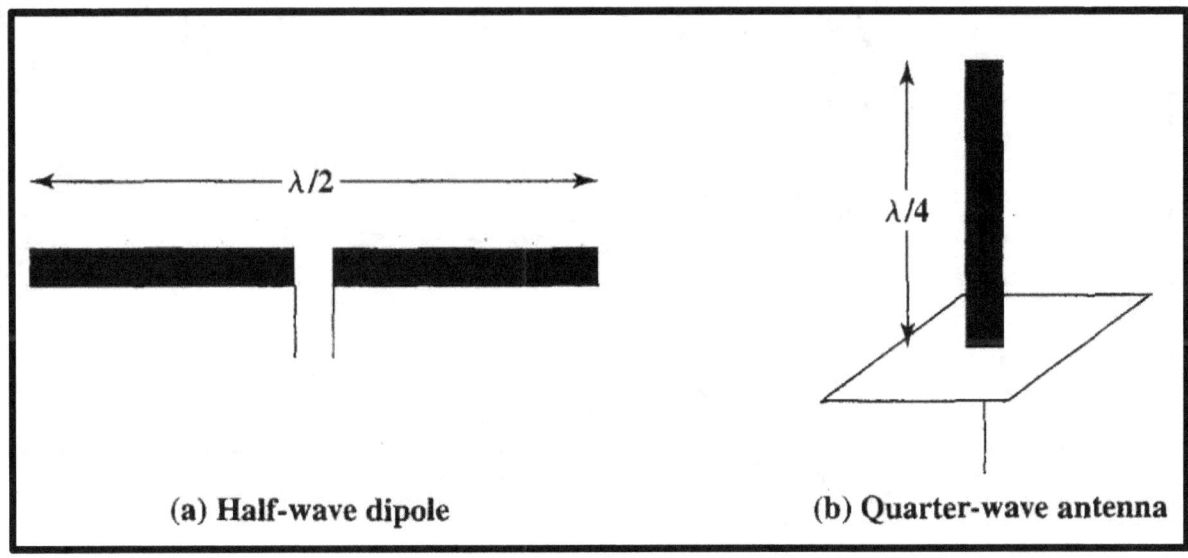

Figure 2.9: Dipole Antennas

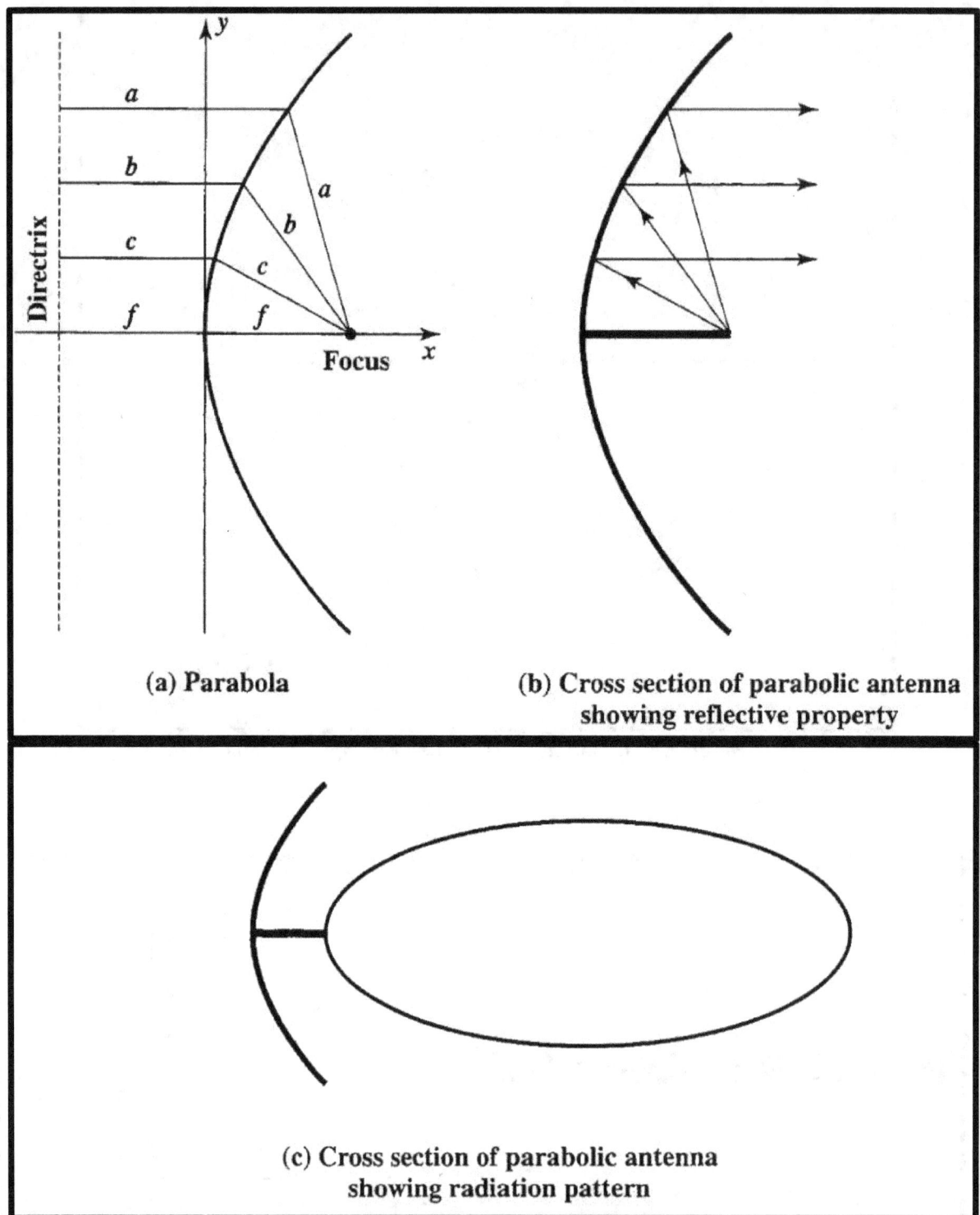

Figure 2.10: Parabolic Reflective Antennas

Antenna Gain:

Antenna Gain is a measure of the directionality of an antenna. Antenna gain is defined as **the power of output, in a particular direction, compared to that produced in any direction by a perfect omnidirectional antenna (isotropic antenna).** For example, if an antenna has a gain of 3 dB, that antenna improves upon the isotropic antenna in that direction by 3 dB. The **increased power radiated in a given direction is at the expense of other directions.** In effect, **increased power is radiated in one direction by reducing the power radiated in other directions.** It is important to note that **antenna gain does not refer to obtaining more output power** than input power **but rather to directionality**.

A concept related to that of antenna gain is the effective area of an antenna. The **effective area of an** antenna is related to the physical size of the antenna and to its shape. The **relationship between antenna gain and effective area** is:

$$G = \frac{4\pi A_e}{\lambda^2} = \frac{4\pi f^2 A_e}{c^2}$$

where

G = antenna gain
A_e = effective area
f = carrier frequency
c = speed of light ($\approx 3 \times 10^8$ m/s)
λ = carrier wavelength

2.3.2 Propagation Modes:

A *signal radiated from an antenna* travels along one of three routes: *Ground Wave*, *Sky Wave*, or *Line of Sight (LOS)*.

Ground Wave Propagation:

Ground wave propagation follows the contour (shape) of the earth and can propagate considerable distances, well over the visual horizon. This effect is found in *frequencies up to about 2 MHz*. Several factors account for the tendency of electromagnetic wave in this frequency band to follow the earth's curvature.

One factor is that the *electromagnetic wave induces a current in the earth's surface*, the result of which is to slow the wave front near the earth, *causing the wave front to tilt downward* and hence *follow the earth's curvature*.

Another factor is **diffraction**, which is a phenomenon having to do with the behavior of electromagnetic waves in the presence of obstacles. *Electromagnetic waves* in this frequency range are *scattered by the atmosphere* in such a way that *they do not penetrate the upper atmosphere*.

The best-known example of ground wave communication is *AM radio*.

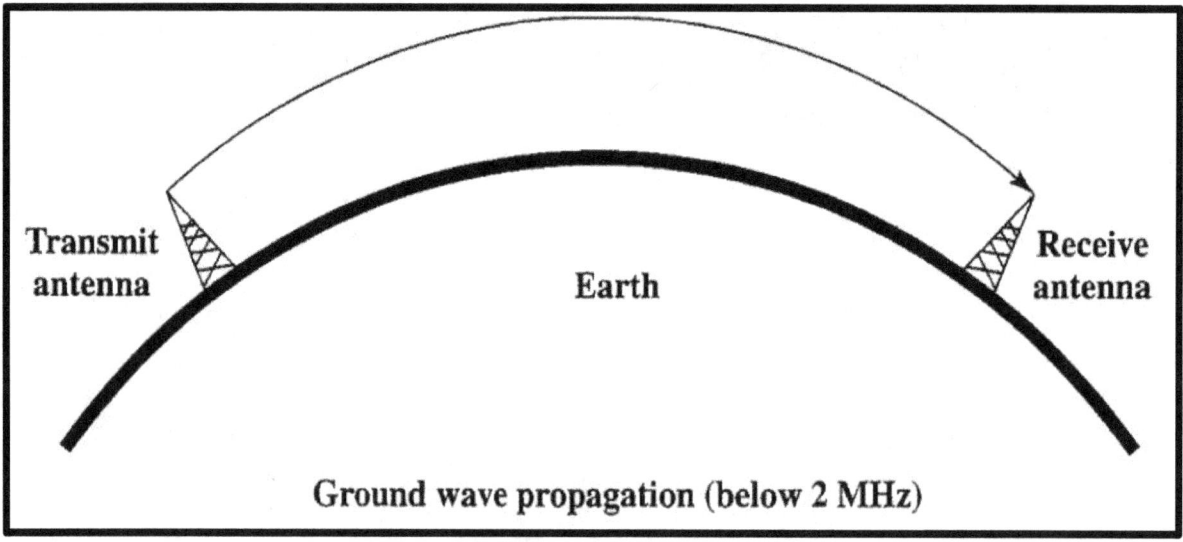

Figure 2.11: Ground Wave Propagation

Sky Wave Propagation:

Sky wave propagation is used for international broadcasts such as BBC and Voice of America. With **sky wave propagation**, a **signal from an earth-based antenna is reflected from the ionosphere** (upper atmosphere on Earth) **back down to earth**.

Although it appears the wave is reflected from the ionosphere as if the ionosphere were a hard-reflecting surface, **the effect is in fact caused by refraction** (bending of signal as it passes from one level of density to another).

A sky wave signal can travel through a number of hops, bouncing back and forth between the ionosphere and the earth's surface. With this propagation mode, a **signal can be picked up thousands of kilometers from the transmitter**.

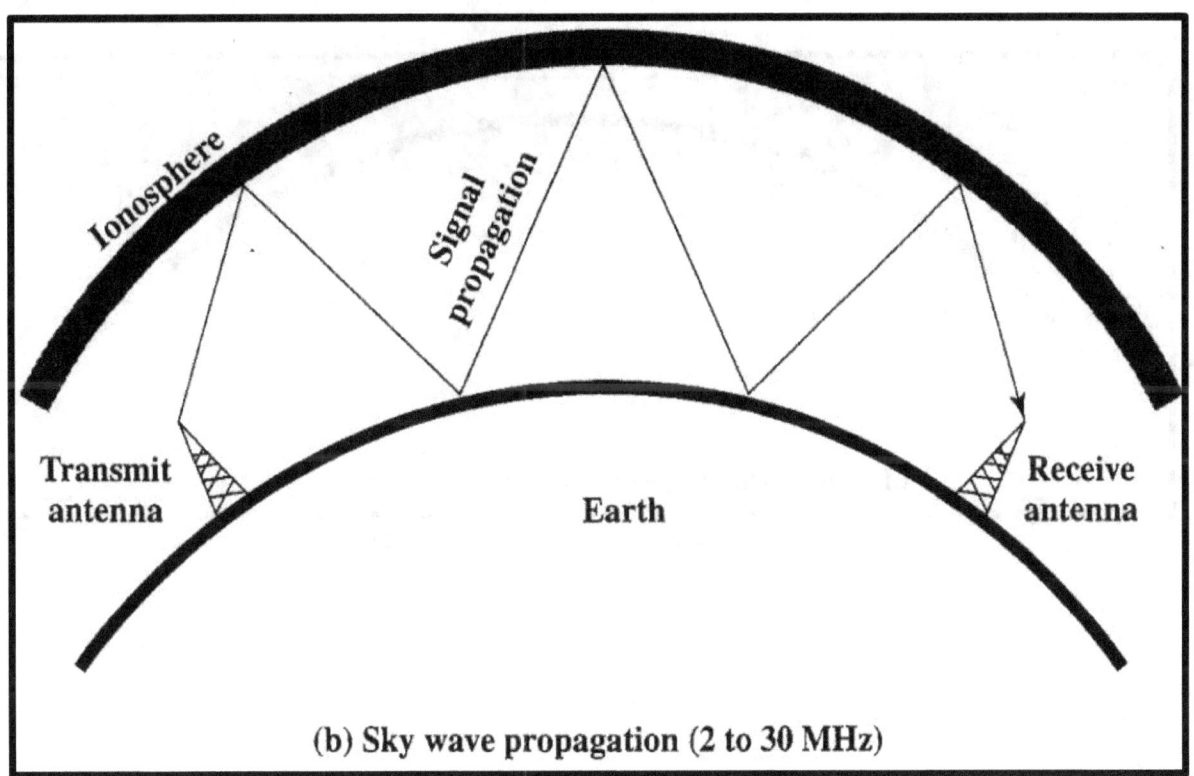

Figure 2.12: Sky Wave Propagation

Line-of-Sight (LOS) Propagation:

Above 30 MHz, neither ground wave nor sky wave propagation modes operate, and communication must be by **line of sight**.

For **satellite communication**, a signal above 30 MHz is not reflected by the ionosphere and therefore can be transmitted between an earth station and a satellite overhead that is not beyond the horizon.

For **ground-based communication**, the **transmitting and receiving antennas** must be **within an effective line of sight of each other**. The term effective is used because microwaves are bent or refracted by the atmosphere. The amount and direction of the bend depends on conditions, but generally microwaves are bent with the curvature of the earth and will therefore propagate farther than the optical line of sight.

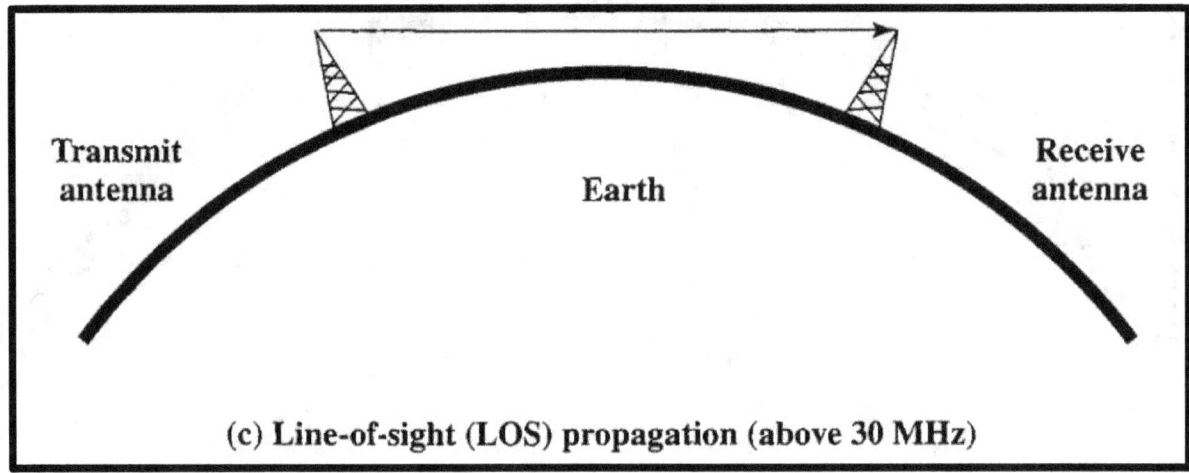

Figure 2.13: Line-of-Sight (LOS) Propagation

2.3.3 Line of Sight Transmission:

In any communication system, the signal that is received will differ from the signal that is transmitted, due to various transmission impairments. The most significant impairments are:.

- Attenuation and Attenuation Distortion
- Free Space Loss
- Noise
- Atmospheric Absorption
- Multipath
- Refraction

Attenuation:

The **strength of a signal falls off with distance** over any transmission medium. For **guided media**, this **reduction in strength**, or **attenuation**, is generally **exponential** and thus is typically expressed as a **constant number of decibels per unit distance**. For **unguided media**, attenuation is a **more complex function of distance and the makeup of the atmosphere**.

Free Space Loss:

For any type of wireless communication, the **signal disperses with distance**. Therefore, an antenna with a fixed area will receive less signal power the farther it is from the transmitting antenna. For **satellite communication**, this is the primary mode of signal loss. Even if no other sources of attenuation or impairment are assumed, a transmitted signal attenuates over distance because the signal is being spread over a larger and larger area. This form of attenuation is known as **Free Space Loss**, which can be express in terms of the **ratio of the radiated power (P_t) to the power received by the antenna (P_r)**.

Noise:

For any data **transmission event**, the **received signal** will consist of the **transmitted signal, modified by the various distortions** imposed by the transmission system, plus additional **unwanted signals** that are inserted somewhere between transmission and reception. These unwanted signals are **referred to as noise**. Noise may be divided into four categories:

- **Thermal Noise**: Thermal noise is due to **thermal agitation of electrons**. It is present in all electronic devices and transmission media and is a **function of temperature**. Thermal noise is uniformly distributed across the frequency spectrum and hence is often referred to as **white noise**.

- ***Intermodulation Noise***: When ***signals at different frequencies share the same transmission medium***, the result may be ***intermodulation noise***. Intermodulation noise produces signals at a frequency that is the sum or difference of the two original frequencies or multiples of those frequencies.

- ***Crosstalk***: Crosstalk has been experienced by anyone who, while using the telephone, has been able to hear another conversation; it is an ***unwanted coupling between signal paths***. It can occur by electrical coupling between nearby twisted pairs or, rarely, coaxial cable lines carrying multiple signals. ***Crosstalk*** can also occur when ***unwanted signals are picked up by microwave antennas***.

- ***Impulse Noise***: Impulse noise consists of ***irregular pulses or noise spikes of short duration*** and of relatively high amplitude. It is generated from a variety of causes, including external ***electromagnetic disturbances, such as lightning***.

Atmospheric Absorption:

An additional loss between the transmitting and receiving antennas is atmospheric absorption. ***Water vapor and oxygen*** contribute most to attenuation. Also, ***rain and fog*** cause scattering of radio waves that results in attenuation. This can be a major cause of signal loss. Thus, in areas of significant precipitation, either ***path lengths have to be kept short*** or ***lower-frequency bands should be used***.

Multipath:

For wireless facilities where there is a relatively free choice of where antennas are to be located, they can be placed so that if there are ***no nearby interfering obstacles***, there is a ***direct line-of-sight path from transmitter to receiver***. This is generally the case for many satellite facilities and for point-to-point microwave.

In other cases, such as ***mobile telephony***, there are ***obstacles in abundance***. The ***signal can be reflected by such obstacles*** so that multiple copies of the signal with varying delays can be received. In extreme cases, the ***receiver may capture only reflected signals*** and not the direct signal. As a result, the composite ***signal can be either larger or smaller than the direct signal***. This is known as ***Multipath***.

Refraction:

Radio waves are ***refracted (or bent)*** when they propagate through the atmosphere. The refraction is caused by ***changes in the speed of the signal with altitude*** or by other spatial changes in the atmospheric conditions. Normally, the speed of the signal increases with altitude, causing ***radio waves to bend downward***.

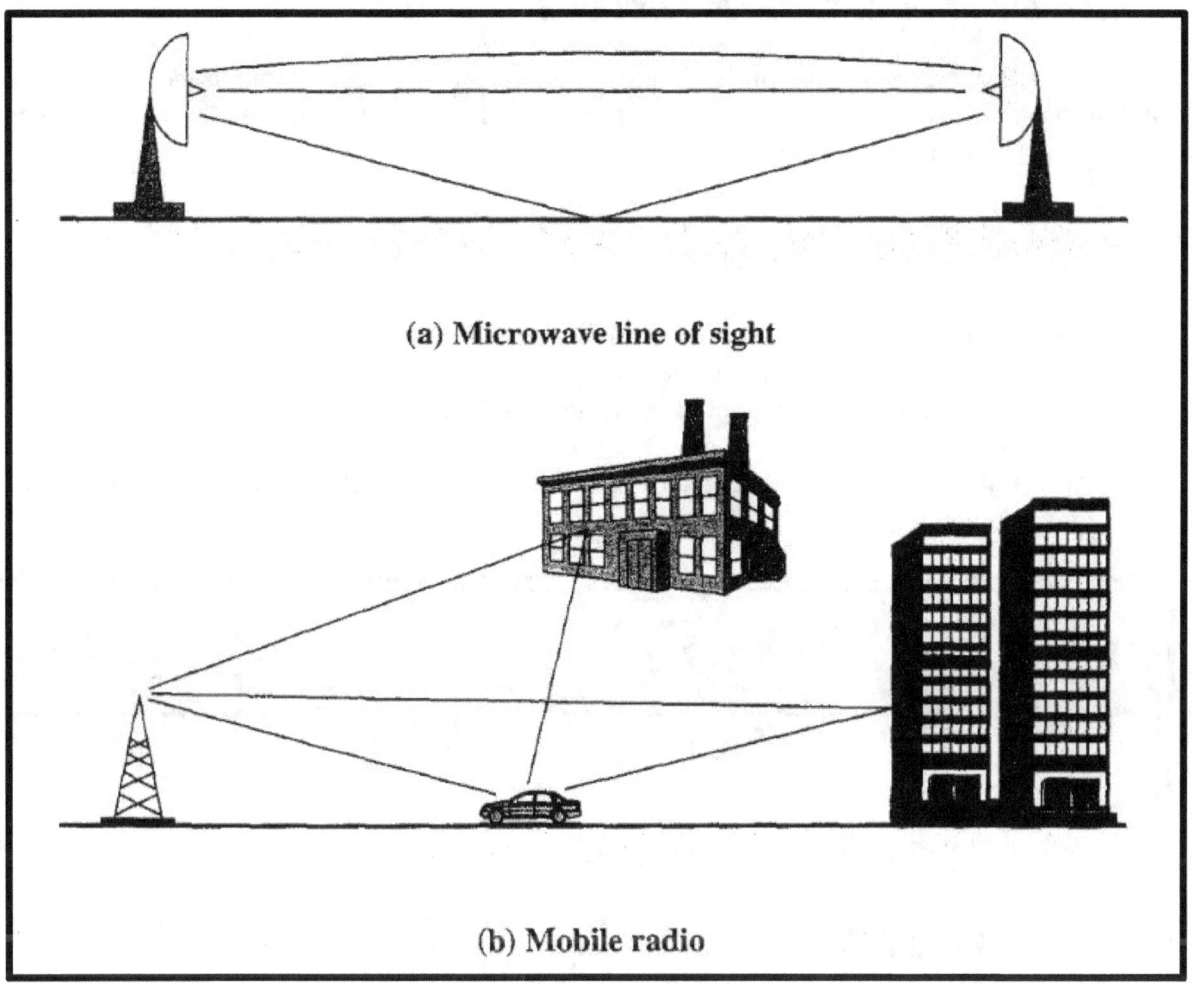

Figure 2.14: Multipath Interference

2.3.4 Fading in the Mobile Environment:

The term *fading*, or, *small-scale fading*, means **rapid fluctuations of the amplitudes, phases, or multipath delays of a radio signal over a short period or short travel distance**. Three propagation mechanisms, *plays a role in fading*:

- o **Reflection** occurs when an electromagnetic signal encounters a surface that is large relative to the wavelength of the signal.

- o **Diffraction** occurs at the edge of an impenetrable body that is large compared to the wavelength of the radio wave. When a radio wave encounters such an edge, waves propagate in different directions with the edge as the source.

- o **Scattering** occurs when the size of an obstacle is on the order of the wavelength of the signal or less than that.

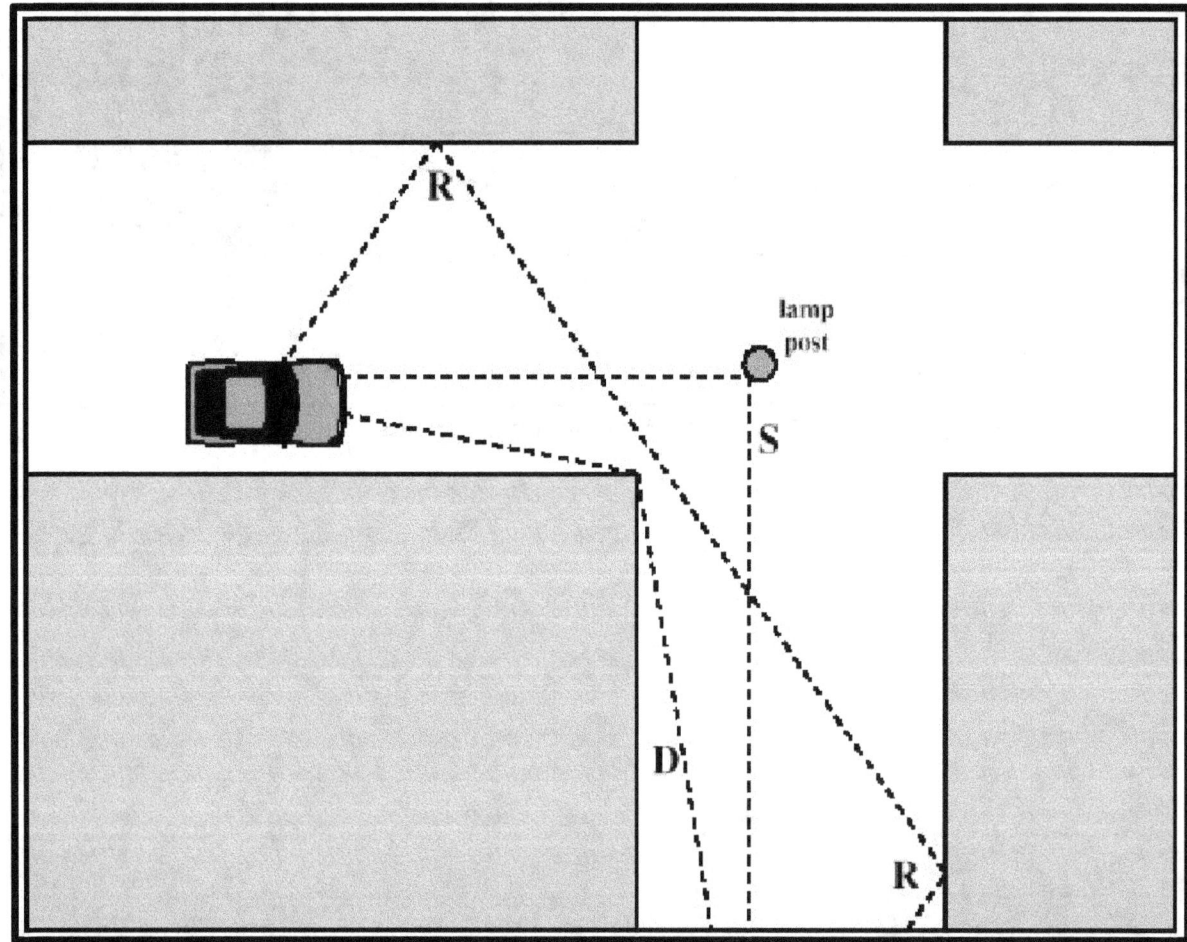

Figure 2.15: Important Propagation Mechanisms: Reflection (R), Scattering (S), Diffraction (D)

Types of fading:

Fading effects in a mobile environment can be classified as either **fast** or **slow**.

Fast fading, as the name suggests, results in **massive change in frequency** over a **short period of time**.

Slow fading results in **moderate change in the frequency** over a period of time.

Fading effects can also be classified as **flat** or **selective**.

Flat fading, or **nonselective fading**, is that type of fading in which **all frequency components** of the received signal **fluctuate in the same proportions** simultaneously.

Selective fading affects the different spectral components of a radio signal in an **unequal manner**.

2.4 Spread Spectrum:

2.4.1 The Concept of Spread Spectrum:

An increasingly important form of communications is known as **spread spectrum**. It can be used to transmit either **analog or digital data**, using an **analog signal**. The essential idea is to **spread the information signal over a wider bandwidth** to make jamming and interception more difficult.

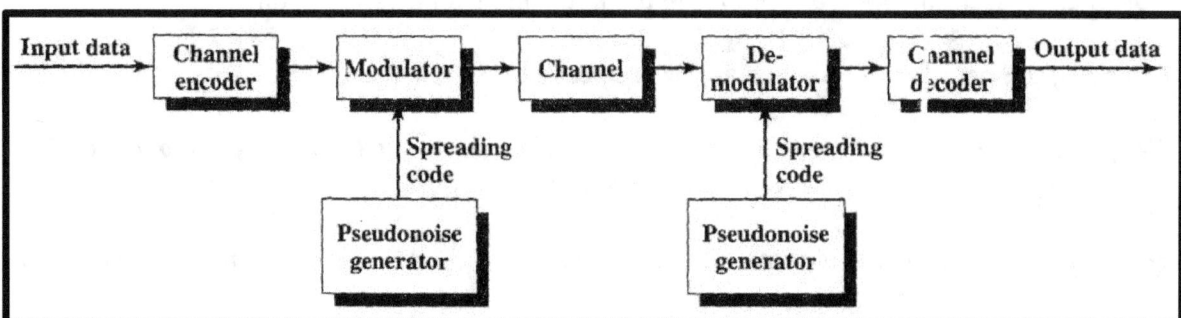

Figure 2.16: Spectral Communication in Wireless Networks

The above figure highlights the key characteristics of any **spread spectrum system**. Input is fed into a **channel encoder** that produces an **analog signal with a relatively narrow bandwidth**. This signal is **further modulated** using a sequence of digits known as a **spreading code**. Typically, but not always, the **spreading code is generated by a pseudo-noise generator**. The effect of this modulation is to significantly **increase the bandwidth** (spread the spectrum) of the signal to be transmitted.

On the **receiving end**, the **same spreading code** is used to **demodulate** the spread spectrum signal. Finally, the signal is fed into a **channel decoder** to **recover the data**.

Several things can be gained from this apparent waste of Spectrum:

- o We can gain immunity from various kinds of noise and multipath distortion.
- o It can also be used for hiding and encrypting signals. Only a recipient who knows the spreading code can recover the encoded information.
- o Several users can independently use the same higher bandwidth with very little interference. This property is used in cellular telephony applications, with a technique known as **Code Division Multiple Access (CDMA)**.

The first type of spread spectrum developed is known as **Frequency Hopping**. A more recent type of spread spectrum is **Direct Sequence**. The next section discusses both of them in detail.

2.4.2 Frequency Hopping Spread Spectrum (FHSS):

With *frequency hopping spread spectrum (FHSS)*, the signal is *broadcast over a random series of radio frequencies, hopping from frequency to frequency at fixed intervals*. A receiver, hopping between frequencies in synchronization with the transmitter, picks up the message. When an attacker attempts to jam the signal on one frequency & succeeds, he might only be able to knock out a few bits of the signal.

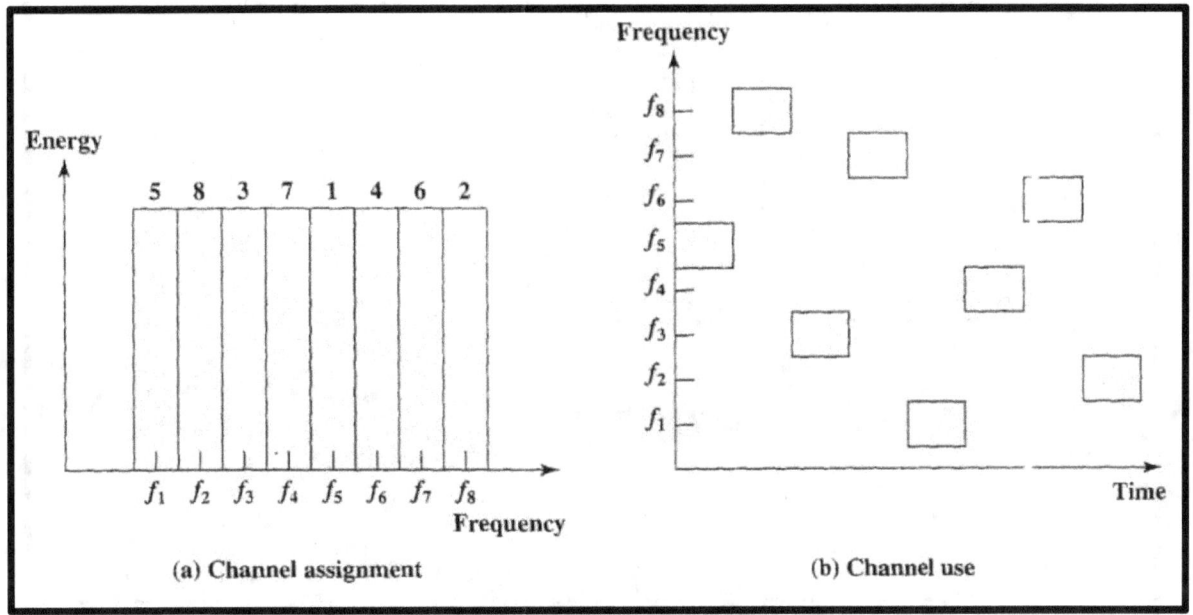

Figure 2.17: Frequency Hopping Example

The above figure shows an example of a frequency hopping signal. *A number of channels are allocated for the FH (Frequency Hopping) signal*. Typically, there are 2^k carrier frequencies forming 2k channels. The spacing between carrier frequencies and the *width of each channel corresponds to the bandwidth of the input signal.*

The *transmitter operates in one channel at a time for a fixed interval*. During that interval, some number of bits are transmitted using some encoding scheme. The *sequence of channels used is dictated by a spreading code*. Both transmitter and receiver use the same code to tune into a sequence of channels in synchronization.

A *typical block diagram for a frequency hopping* is shown in figure 18. For transmission, *binary data are fed into a modulator* using some *digital-to-analog encoding* scheme, such as *Frequency-Shift Keying (FSK)* or *Binary Phase-Shift Keying (BPSK)*. A *pseudo-noise (PN) source* serves as an *index into a table of frequencies*; this is the *spreading code*. Each k bits of the pseudo-noise (PN) source specifies one of the 2^k carrier frequencies. At *each successive interval, a new carrier frequency c(t) is selected. This frequency is then modulated by the signal produced from the initial modulator* to produce a new signal set with the same shape but now centered on the selected carrier frequency. On **reception**, the spread spectrum signal is **demodulated using the same sequence of PN-derived frequencies** and then **demodulated to produce the output data**.

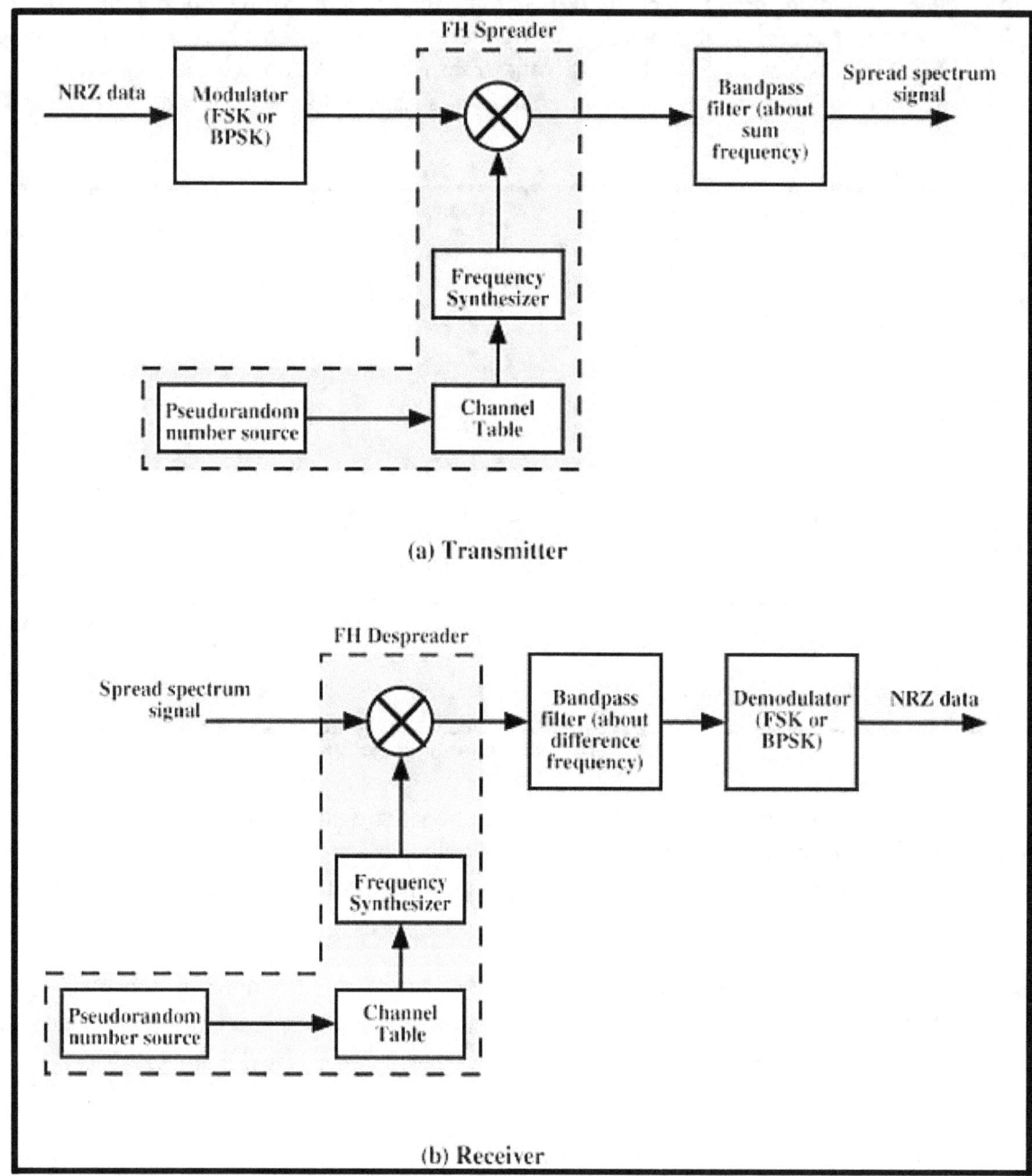

Figure 2.18: Frequency Hopping Spread Spectrum System (Block Diagram)

2.4.3 Direct Sequence Spread Spectrum (DSSS):

For *direct sequence spread spectrum (DSSS)*, *each bit in the original signal* is represented by *multiple bits in the transmitted signal*, using a *spreading code*. The *spreading code* spreads the signal across a wider frequency band in direct proportion to the number of bits used. That is, *a n-bit spreading code spreads the signal across a frequency band that is n times greater than a 1-bit spreading code*.

One technique for *direct sequence spread spectrum* is to combine the digital information stream with the spreading code bitstream using an *Elusive-OR (XOR)*. The *XOR* obeys the following rules:

(0,0) = 0   (0,1) = 1   (1,0) = 1   (1,1) = 0

Following image shows an *example of DSSS*:

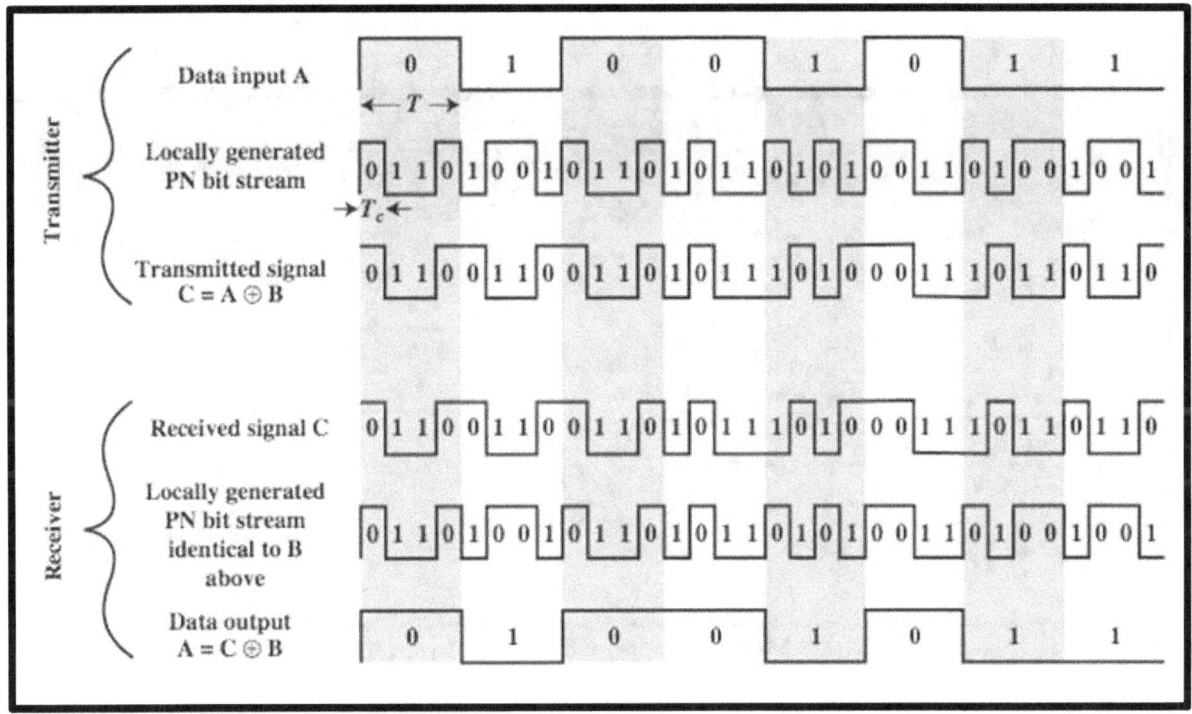

Figure 2.19: Direct Sequence Spread Spectrum System

With *DSSS*, the *original signal is modulated with a bit sequence known as a Pseudo Noise (PN) code*; this *PN code* consists of a *radio pulse* that is *much shorter in duration (larger bandwidth) than the original signal*. This *modulation* of the original signal scrambles and spreads the pieces of data, and thereby *resulting in a bandwidth size nearly identical to that of the PN sequence*. More *bandwidth modulated* to the original signal results in *better resistance against interference*.

Following figure shows the *block diagram of Direct Sequence Spread Spectrum System (DSSS)*:

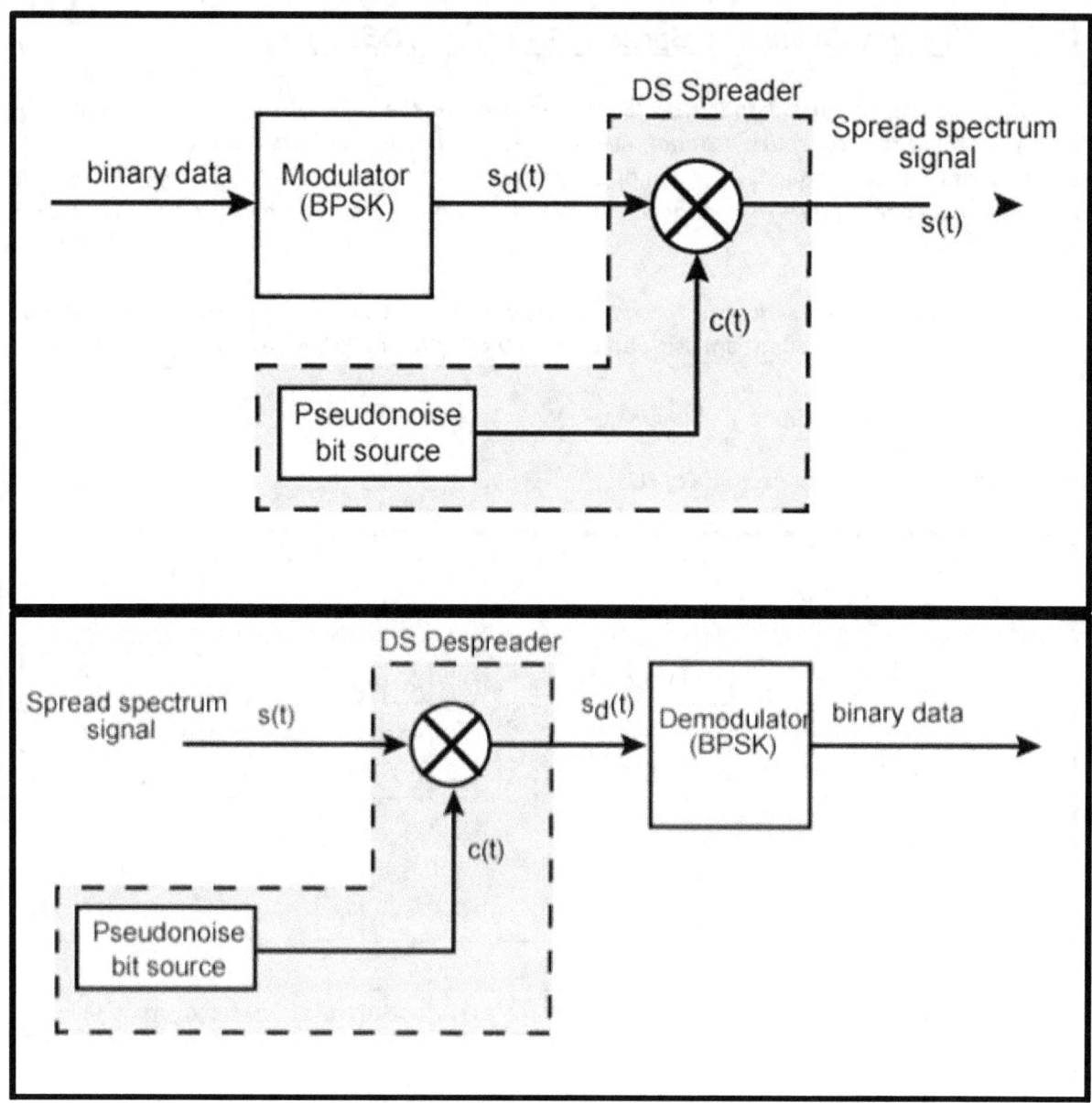

Figure 2.20: Direct Sequence Spread Spectrum System (Block Diagram)

2.5 Coding & Error Control:

Regardless of the design of the transmission system, there will be errors, resulting in the change of one or more bits in a transmitted signal. **Three approaches** are in common use for **coping with data transmission errors**:

- *Error Detection Codes*
- *Error Correction Codes or Forward Error Correction (FEC) Codes*
- *Automatic repeat request (ARQ) protocols*

An **error detection code** simply detects the presence of an error. Such codes are used in conjunction with a protocol at the data link or transport level that uses an ARQ scheme.

With an **ARQ scheme**, a receiver discards a block of data in which an error is detected and the transmitter retransmits that block of data.

FEC codes are designed not just to detect but correct errors, avoiding the need for retransmission. FEC schemes are frequently used in wireless transmission, where retransmission schemes are highly inefficient.

2.5.1 Error Detection:

Following are thee commonly used error detection techniques:

- **Parity Check**:

    The simplest error detection scheme is to **append a parity bit to the end of a block of data**. A typical example is character transmission, in which a parity bit is attached to each 7-bit character. The value of this bit is selected so that the character has an even number of Is (even parity) or an odd number of Is (odd parity). The use of the parity bit is not full-proof, as noise impulses are often long enough to destroy more than one bit.

- **Cyclic Redundancy Check (CRC)**:

    One of the most common, and one of the most powerful, error-detecting codes is the cyclic redundancy check (CRC), which can be described as follows. Given a *k bit block of message*, the transmitter generates an *(n - k)-bit sequence*, known as a *frame check sequence (FCS)*, such that *the resulting frame, consisting of n bits*, is **exactly divisible** by some predetermined number. The **receiver** then **divides** the incoming frame by that number and, if there is **no remainder**, assumes there was **no error**.

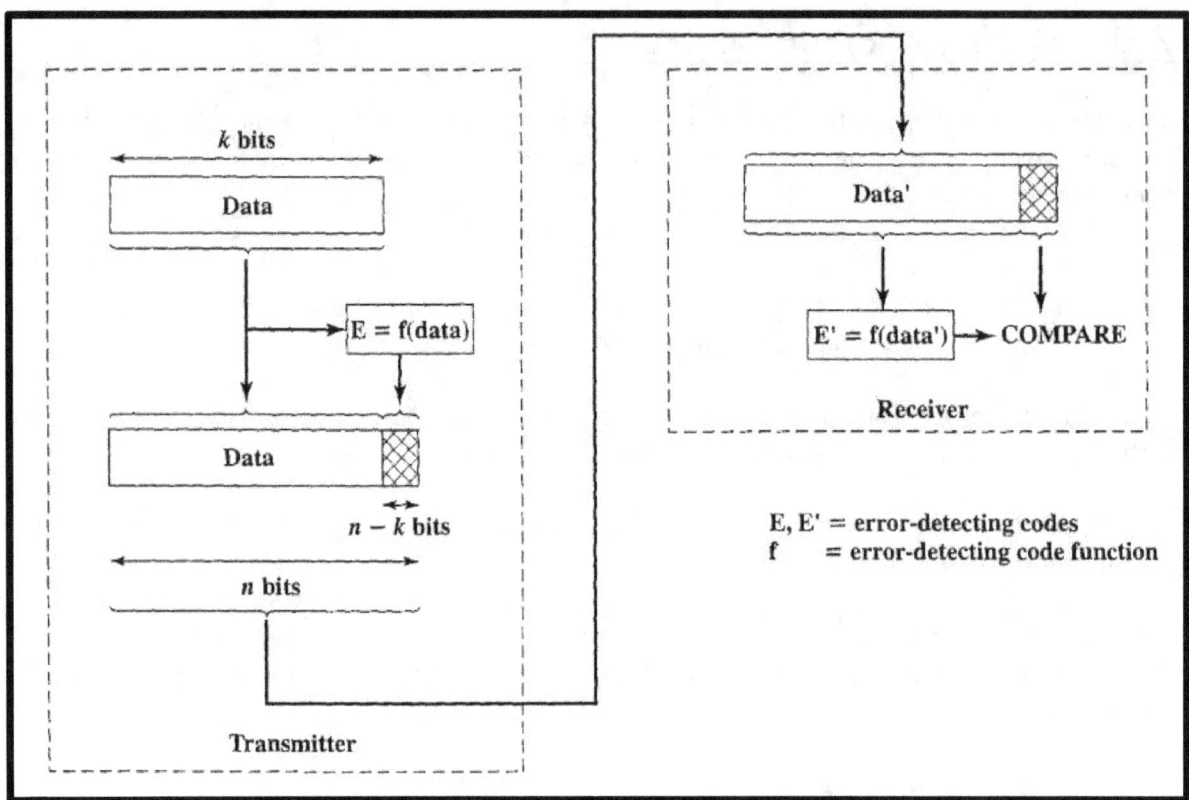

Figure 2.21: Error Detection Process

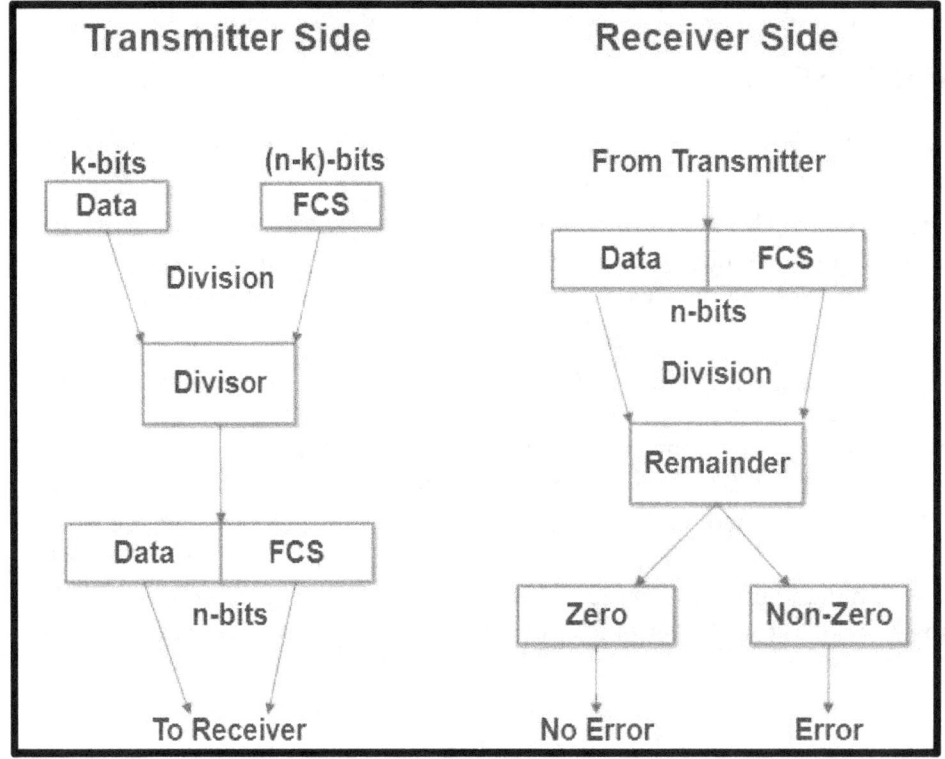

Figure 2.22: CRC Error Detection

2.5.2 Block Error Correction Code:

Instead of retransmitting the signal in case of an error, it would be desirable to enable the receiver to *correct errors* in an incoming transmission on the basis of the bits in that transmission. The below figure shows how it's done:

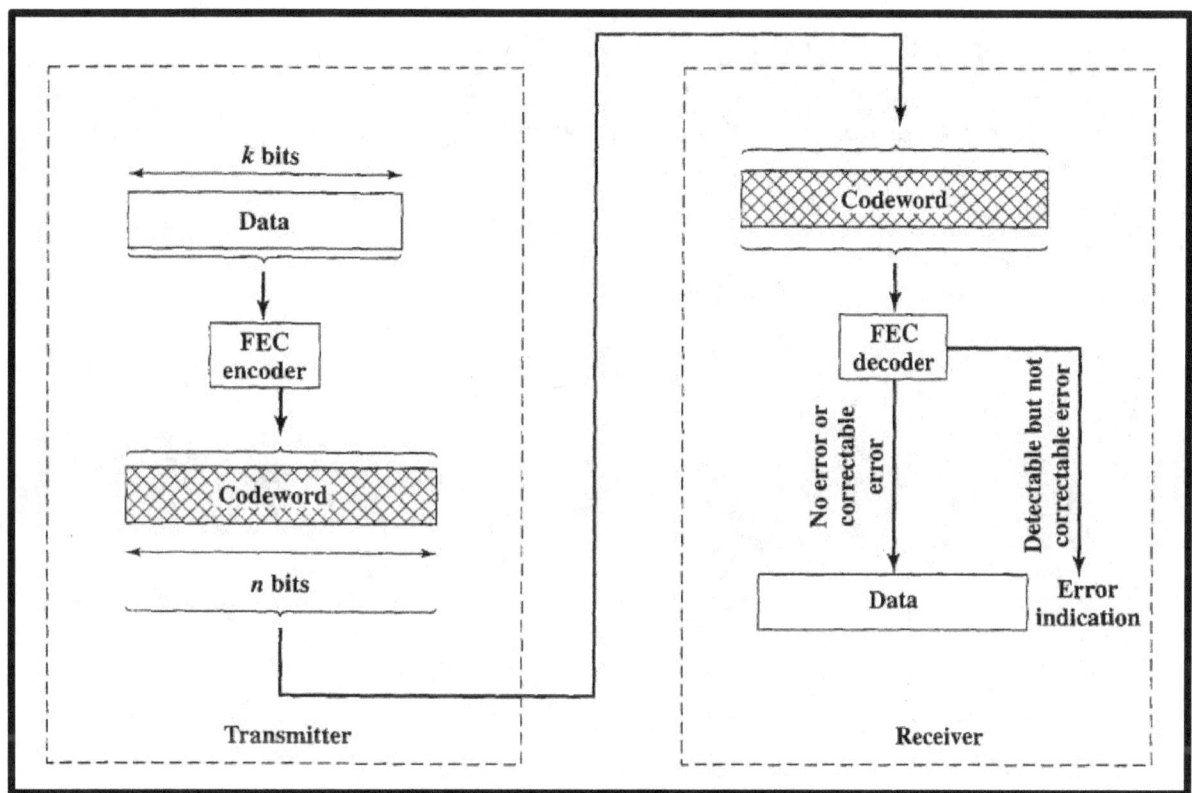

Figure 2.23: Forward Error Correction

On the transmission end, *each k-bit block of data is mapped into an n-bit block (n > k) called a codeword*, using an FEC (forward error correction) Encoder. The *codeword is then transmitted*.

At the *receiver end*, the incoming signal is *demodulated* to produce a *bit string* that is *similar to the original codeword but may contain errors*. This block is passed through an *FEC decoder*, with one of *four possible outcomes*:

1. If there are **no bit errors**, the input to the FEC decoder is identical to the original codeword, and the decoder produces the original data block as output.

2. For certain error patterns, it is possible for the decoder to **detect and correct those errors**. Thus, even though the incoming data block differs from the transmitted codeword, the FEC decoder is able to map this block into the original data block.

3. For certain error patterns, the decoder can **_detect but not correct the errors_**. In this case, the decoder simply reports an uncorrectable error.
4. For certain, typically rare, error patterns, the decoder **_does not detect that any errors have occurred_** and maps the incoming n-bit data block into a k-bit block that differs from the original k-bit block.

How is it possible for the decoder to correct bit errors?

In essence, error correction works by **adding redundancy to the transmitted message**. The redundancy makes it possible for the receiver to deduce what the original message was, even in the face of a certain level of error rate.

2.5.3 Automatic Repeat Request (ARQ):

Automatic Repeat request (ARQ) is an **error-control method** for data transmission that uses **acknowledgements & timeouts** to achieve reliable data transmission over an unreliable service. If the **sender does not receive an acknowledgment before the timeout**, it usually **re-transmits** the frame/packet **until the sender receives an acknowledgment or exceeds a predefined number of re-transmissions (until timeout)**.

The types of ARQ protocols include **Stop-and-wait ARQ, Go-Back-N ARQ,** and **Selective Repeat ARQ**. All three protocols usually use some form of **sliding window protocol** to tell the transmitter to determine which (if any) packets need to be retransmitted. These protocols reside in **the Data Link or Transport Layers of the OSI model**.

Chapter 3

Introduction to GSM & GPRS

3.1 Multiple access in Wireless System:

3.1.1 Multiple Access Scheme:

When nodes or stations use a **common radio channel** for communication, called a **multipoint or broadcast channel**, we need a **multiple-access protocol** to coordinate access to the link. This is because we need to control simultaneous access of radio channel in order to **avoid collisions**.

In a connection-oriented communication, **collisions are undesirable**. For effective utilization, channel has to be utilized intelligently by the use of **multiplexing**. Four major types of **multiple access procedures** are:

- *Frequency Division Multiple Access (FDMA)*
- *Time Division Multiple Access (TDMA)*
- *Code Division Multiple Access (CDMA)*
- *Space Division Multiple Access (SDMA)*

Each of these four procedures are discussed in detail in the following section.

3.1.2 Frequency Division Multiple Access (FDMA):

In *frequency-division multiple access (FDMA)*, the **available bandwidth is divided into frequency bands** & that **each station communicates on a different bandwidth**. In other words, each frequency band is reserved for a specific station, and it belongs to that particular station all the time.

To **prevent station interferences**, the **allocated bands** are **separated** from one another by small **guard band**.

FDMA specifies a **predetermined frequency** band **for the entire period** of communication. This means that **stream data** (a continuous flow of data that may not be packetized) **can easily be used with FDMA**. This feature can be used in cellular telephone systems. In fact, **FDMA** was indeed used in all the **first-generation analog mobile networks** like **TACS (Total Access Communication System) in UK** and **AMPS *(Advanced Mobile Phone System) in USA***.

FDMA is **compatible** with **both digital and analog** signals. It is **devoid of timing issues** that exist in TDMA. However, one **disadvantage of FDMA is crosstalk**, which can cause interference between frequencies and interrupt the transmission

We need to emphasize that although **FDMA & FDM** conceptually seem similar, there are **differences between them** as follows:

- **FDM (Frequency Division Multiplexing)** is a physical layer technique that combines the signals from low-bandwidth channels and transmits them by using a single high-bandwidth channel. The channels that are combined are low-pass. The FDM multiplexer modulates the signals, combines them, and creates a single bandpass signal.

- **FDMA**, on the other hand, is an **access method** in the data link layer that is used to access channel.

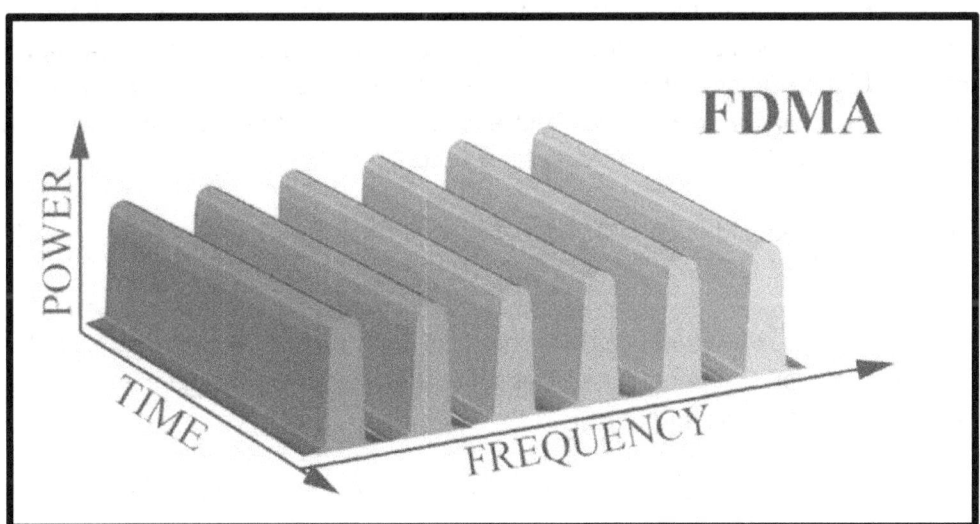

Figure 3.1: Frequency Division Multiple Access (FDMA)

3.1.3   Time Division Multiple Access (TDMA):

In *Time Division Multiple Access (TDMA)*, *several stations* share the *same frequency* bandwidth of the channel on the *basis of time. Each station is allocated a time slot during which it can send data*. Each *station transmits its data in its assigned time slot*. Following image shows the *idea of TDMA*:

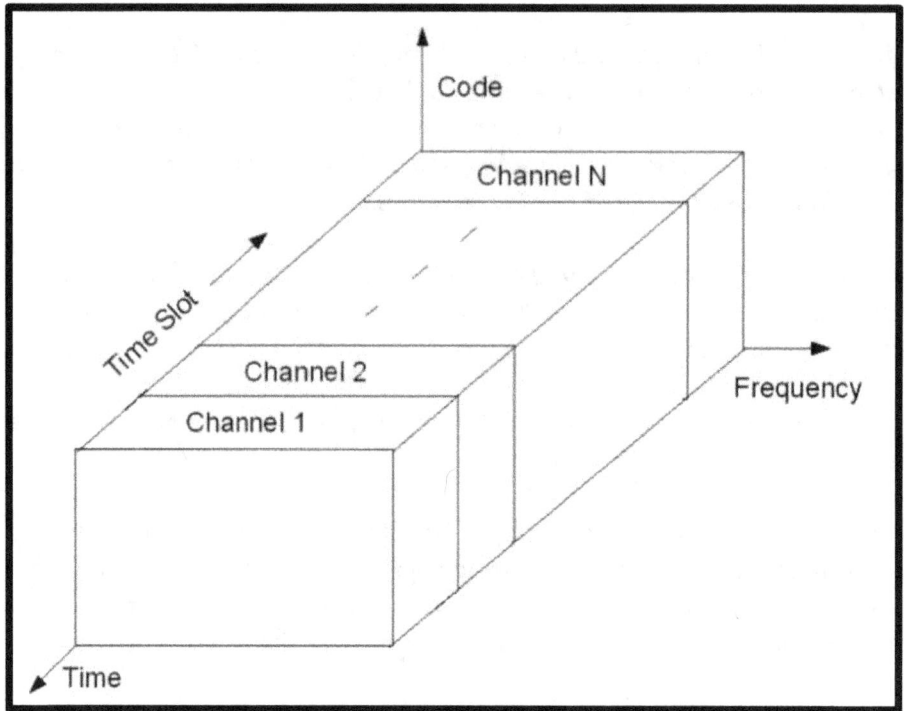

Figure 3.2: Time Division Multiple Access (TDMA)

The *main problem* with TDMA lies in *achieving synchronization between the different stations*. Each station needs to know the beginning of its slot and the location of its slot. This may be difficult because of *propagation delays* introduced in the system if the *stations are spread over a large area*. *For synchronization purpose*, the transmitter and the receiver need to use a *global clock*. And *to compensate for the delays*, we can insert *guard times*.

We need to emphasize that although T*DMA & TDM* conceptually seem similar, there are *differences between them* as follows:

- *TDM (Time Division Multiplexing)* is a physical layer technique that combines the signals from multiple slower channels and transmits them by using a single faster channel. The process uses a physical multiplexer.

- *TDMA*, on the other hand, is an *access method* in the data link layer that tells its physical layer to use the allocated time slot.

3.1.4 Code Division Multiple Access (CDMA):

In *Code Division Multiple Access (CDMA), one channel carries all the signals simultaneously*.

- o *CDMA* differs from FDMA because *only one channel occupies the entire bandwidth* of the link.
- o *CDMA* differs from TDMA because *all stations can send data simultaneously*; there is *no timesharing*.

In CDMA, *each subscriber uses the whole system bandwidth & time*. To permit this without undue interference between the users, CDMA employs *spread spectrum technology* and a *special coding scheme* where *each transmitter is assigned an orthogonal code called chip*.

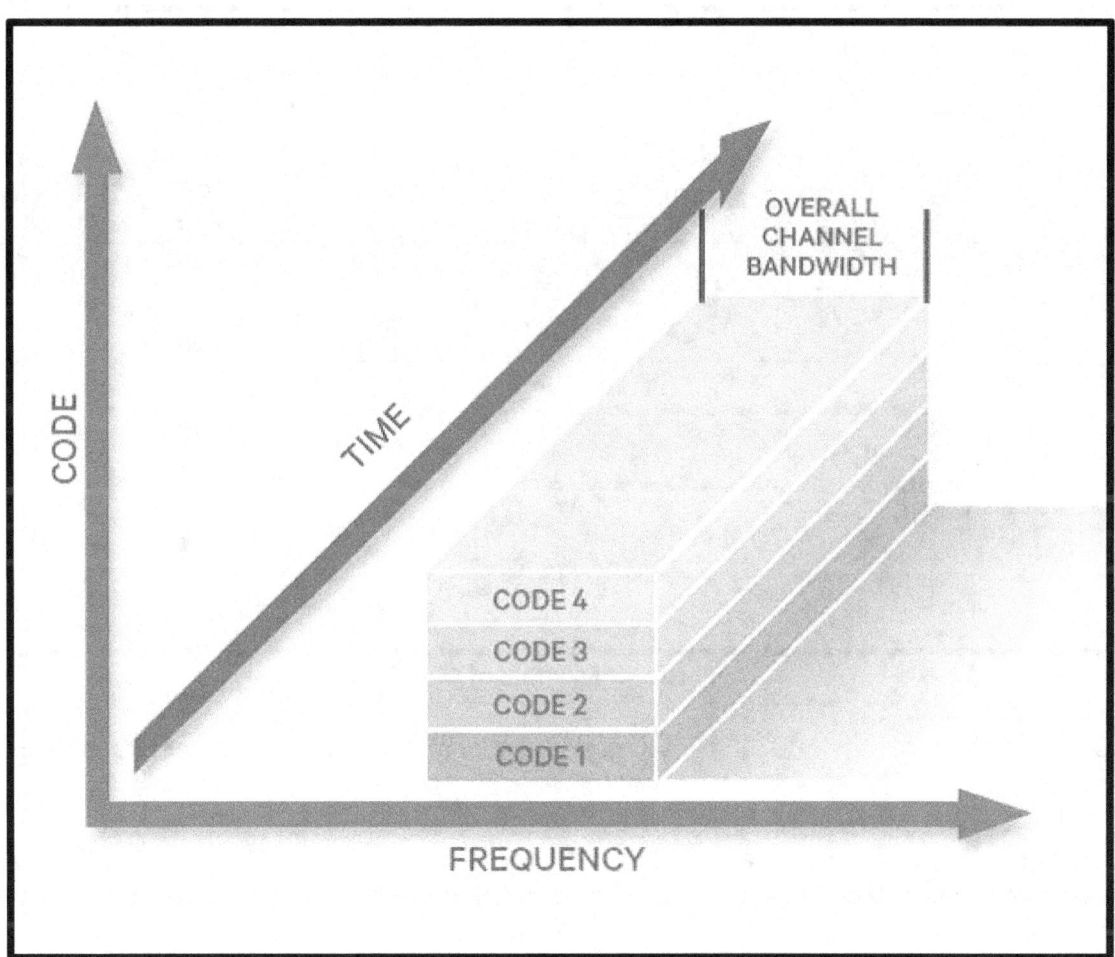

Figure 3.3: Code Division Multiple Access (CDMA)

CDMA simply means *communication with different codes*. For example, in a large room with many people, *two people can talk in English if nobody else understands English*. Another *two people can*

talk in Chinese if they are the only ones who understand Chinese. In other words, the ***common channel*** can easily allow ***communication between several couples***, but in ***different codes***.
Let us assume we have four stations 1, 2, 3, and 4 connected to the same channel.

- o   The data from station 1 is d1, from station 2 is d2, and so on.
- o   The code assigned to the first station is c1, to the second station is c2, and so on.
- o   We assume that the assigned codes have the following two properties:
    - o   If we multiply each code by another code, we get 0.
    - o   If we multiply each code by itself, we get 4 (the number of stations).

With these two properties in mind, let us see how these four stations can send data using the same common channel.

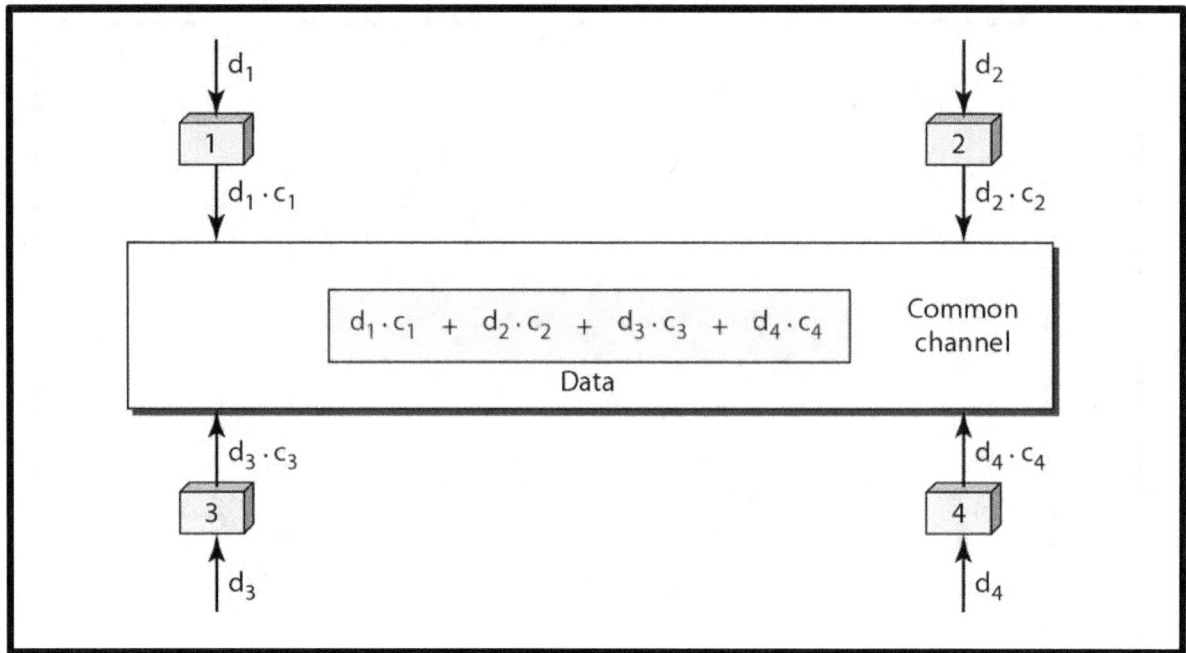

Figure 3.4: Example of CDMA

As shown in Figure 4, station 1 multiplies (a special kind of multiplication) its data by its code to get d1*c1, Station 2 multiplies its data by its code to get d2*c2 and so on.

The data that go on the channel are the sum of all these terms (d1*c1, d2*c2, d3*c3, d4*c4). Any station that wants to receive data from one of the other three multiplies the data on the channel by the code of the sender.

For example, suppose stations 1 and 2 are talking to each other & station 2 wants to hear what station 1 is saying, it multiplies the data on the channel by c1, the code of station 1.

Because (c1*c1) is 4, but (c1*c2), (c1*c3), and (c1*c4) are all 0s, station 2 divides the result by 4 to get the data from station 1.

3.1.5 Space Division Multiple Access (SDMA):

Space Division Multiple Access (SDMA) utilizes the **spatial separation** of the users in order to optimize the use of the frequency spectrum. SDMA is used mostly in wireless and satellite communication. A primitive form of SDMA is when the **same frequency is reused** in **different cells** that are **separated by some distance** in a **cellular wireless network**.

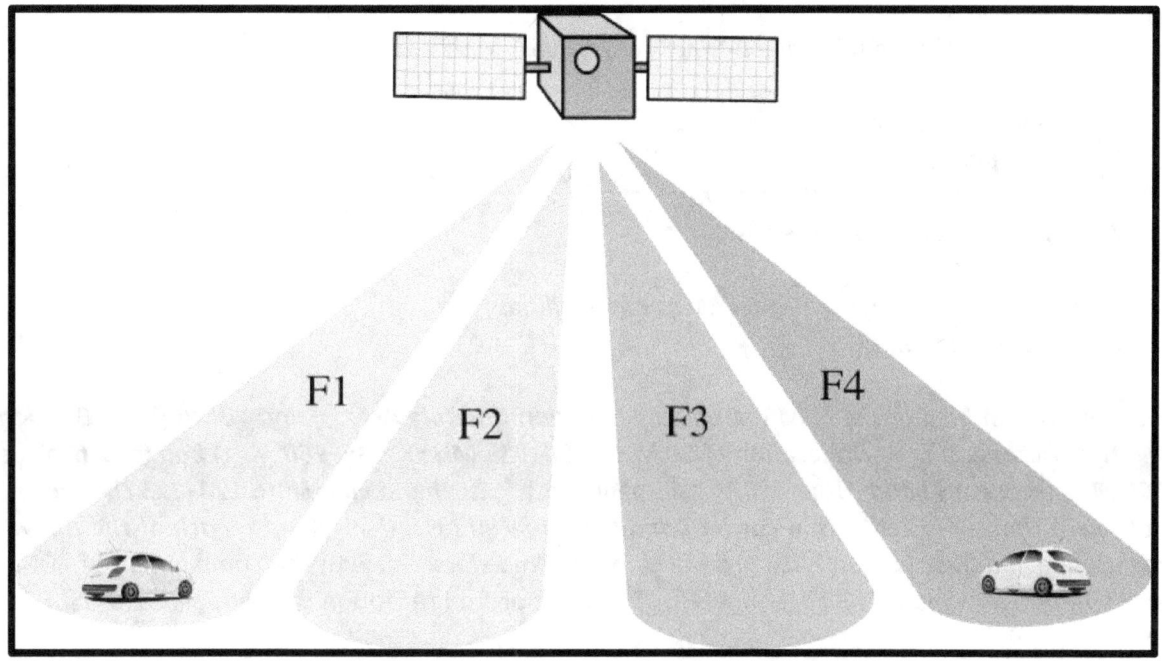

Figure 3.5: Space Division Multiple Access

More recent use of SDMA is observed in **satellite communications**. **Satellite bandwidth** is one of the resource which is **very scarce** and depends on number of transponders in a satellite. Another resource on satellite is **antenna** which **needs to be effectively used**.

To optimize the use of radio spectrum, **directional properties of antennas** are utilized. Satellites use a concept called **polarization**. If satellite will have **two antennas** one **vertically polarized** and the other **horizontally polarized** than two electro-magnetic signals with same frequency can be transmitted to the same satellite. This concept of **frequency re-use** helps in effective use of **satellite bandwidth**.

To serve different users, such satellites also use **directional spot-beam antennas** which makes it **possible for the base station** in SDMA to **track a moving user**.

3.2 Global System for Mobile Communication (GSM):

3.2.1 Introduction to GSM:

GSM stands for Global System for Mobiles. This is a **world-wide standard** for Digital **Cellular Telephony**. GSM was created by the Europeans, and **originally meant Groupe Special Mobile**.

It was meant to fulfill the following **business objectives**:

1. Support for international roaming
2. Good speech quality
3. Ability to support handheld terminals
4. Low terminal and service cost
5. Spectral efficiency
6. Support for a range of new services and facilities
7. ISDN compatibility

GSM uses a **combination of FDMA (Frequency Division Multiple Access) and TDMA (Time Division Multiple Access).** It has an allocation of **50 MHz** (**890–915 MHz** & **935–960 MHz**) **bandwidth in the 900 MHz frequency band**. Using FDMA, this bandwidth is **further divided into 124 (125 channels, 1 not used) channels** each with a **carrier bandwidth of 200 KHz**. Using TDMA, **each of the above-mentioned channels** is then further **divided into 8 time slots**. So, with the combination of FDMA and TDMA, a **maximum of 992 channels** for transmission and reception can be realized in GSM.

In order to serve hundreds of thousands of users, **frequency reuse** is necessary in GSM. This is done **using cells** (discussed in the following sections).

3.2.2 GSM Architecture:

GSM system consists at the minimum **one administrative region** assigned to **one MSC (Mobile Switching Centre)**. **Administrative region** is commonly known as **PLMN (Public Land Mobile Network)**.

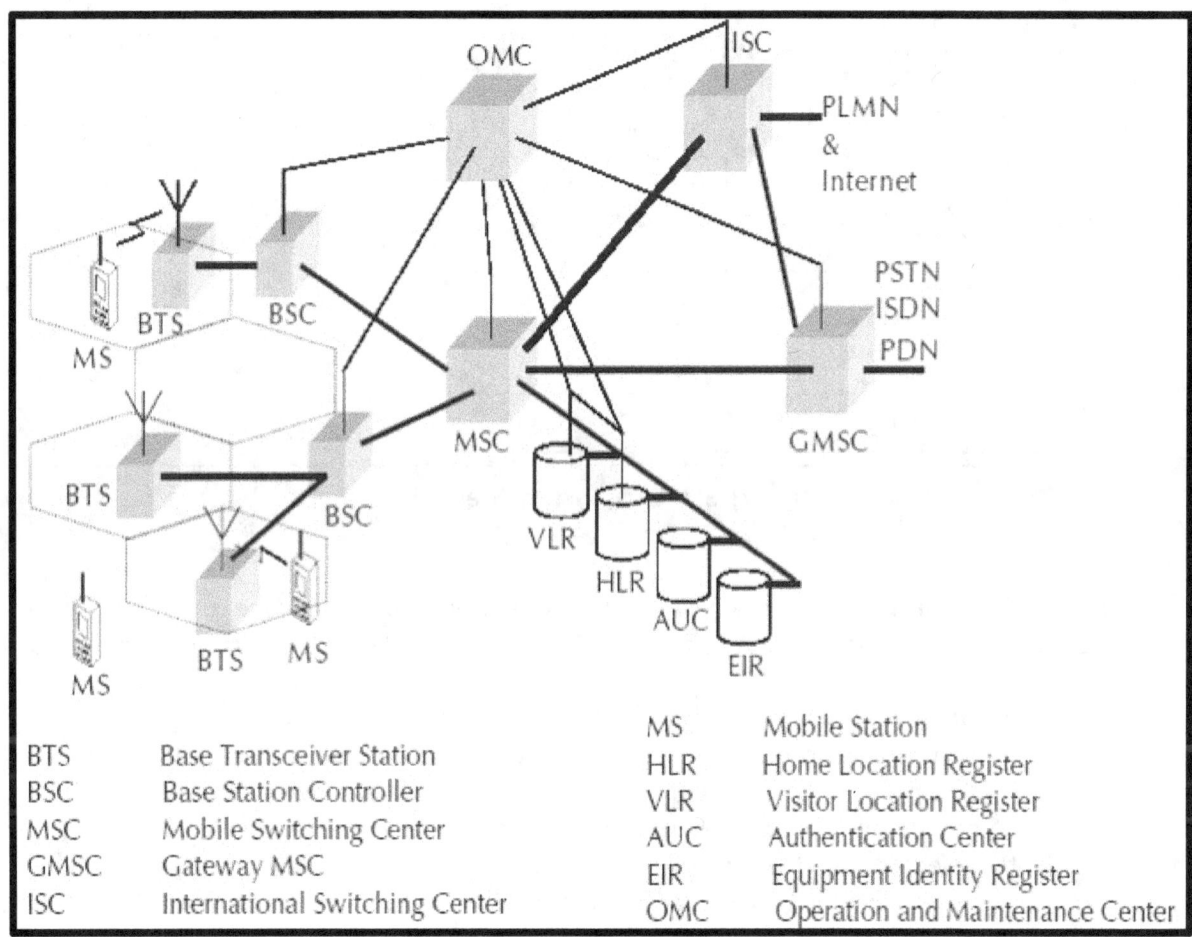

Figure 3.6: GSM Architecture

Each **administrative region** is subdivided into one or many **Location Area (LA)**. One **Location Area (LA)** consists of many **cell groups** and each cell group is assigned to one **BSC (Base Station Controller)**. For each **Location Area (LA)**, there will be at least one **BSC (Base Station Controller)**.

Cells are formed by **radio areas covered by BTS (Base Transceiver Station)**. Several BTSs (Base Transceiver Stations) are **controlled by one BSC (Base Station Controller)**.

Traffic from the **MS (Mobile Station)** is routed through **MSC (Mobile Switching Center)**. Calls originating from or terminating in a fixed network or other mobile networks is handled by the **GMSC (Gateway MSC)**.

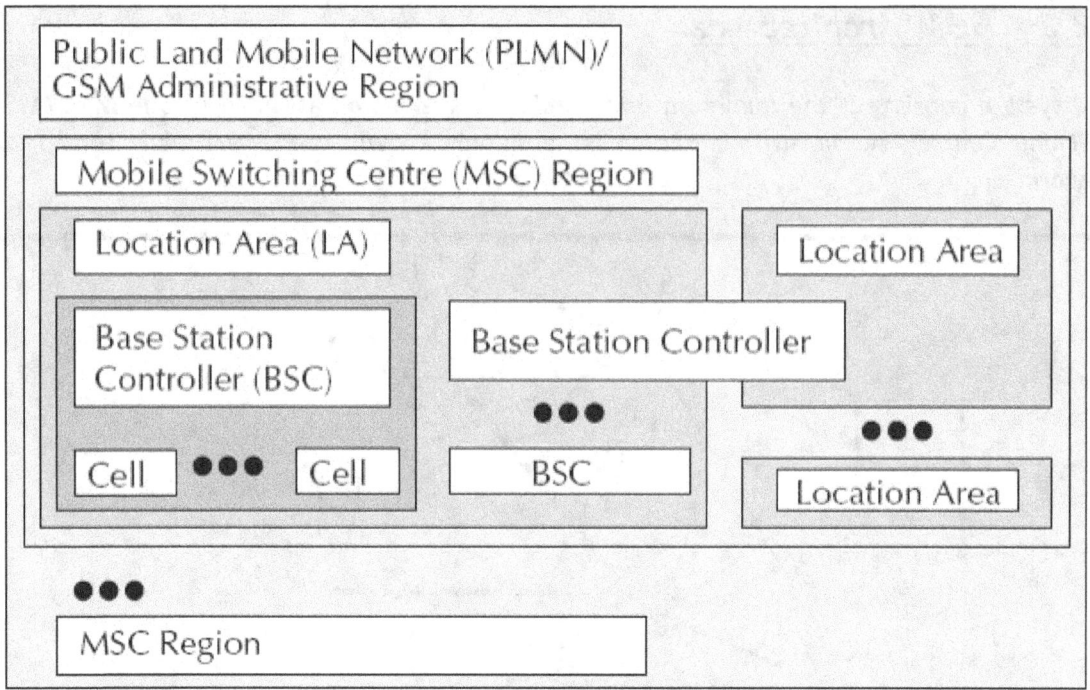

Figure 3.7: GSM System Hierarchy

For **all subscribers registered** with a cellular network operator, **permanent data** such as the service profile is **stored in** the **Home Location Register (HLR)**. The data relates to the following information:

- Authentication information like *IMSI (International Mobile Subscriber Identity)*
- Identification information like *name, address, etc. of the subscriber*
- Identification information like *Mobile Subscriber ISDN (telephone number)*
- Billing Information like *prepaid or postpaid* customer
- Operator *selected denial of service* to a subscriber
- Handling of supplementary services like for *CFU (Call Forwarding Unconditional), CFB (Call Forwarding Busy), CFNR (Call Forwarding Not Reachable) or CFNA (Call Forwarding Not Answered)*
- Storage of *SMS Service Center (SC) number* in case the mobile is not connectable so that whenever the mobile is connectable, a paging signal is sent to the SC
- Provisioning information like whether *long distance and international calls allowed or not*
- Provisioning information like whether *roaming is enabled or not*
- Information related to auxiliary services like *Voice mail, data services, fax services, etc.*
- Information related to auxiliary services like *CLI (Caller Line Identification), etc.*
- Information related to *supplementary services for call routing*. In GSM network, one can customize the personal profile to the extent that while the subscriber is roaming in a foreign PLMN, incoming calls can be blocked. Also, outgoing international calls can be blocked.
- Some variable information like *pointer to the VLR, location area of the subscriber, Power OFF status of the handset, etc.*

Entities in GSM:

1. **The Mobile Station (MS)** - This includes the Mobile Equipment (ME) and the Subscriber Identity Module (SIM).

2. **The Base Station Subsystem (BSS)** - This includes the Base Transceiver Station (BTS) and the Base Station Controller (BSC).

3. **The Network and Switching Subsystem (NSS)** - This includes Mobile Switching Center (MSC), Home Location Register (HLR), Visitor Location Register (VLR), Equipment Identity Register (EIR), and the Authentication Center (AUC).

4. **The Operation and Support Subsystem (OSS)** - This includes the Operation and Maintenance Center (OMC).

5. **Message Center** – Short Message Service (SMS) is one of the most popular services in GSM. SMS is a data service & it allows user to enter text message up to 160 characters in length. SMS is a proactive bearer & is always ON network. Message Center is also referred to as Service Center (SC) or SMS Controller (SMSC). SMSC is a system in core GSM which works as a store & forward system for SMS.

3.2.3 Frequency Reuse in GSM:

To serve hundreds of thousands of users, **the frequency must be reused** and this is **done through cells**. The **area to be covered** is **subdivided into radio zones or cells**. Though in reality these **cells** could be of any shape, for convenient modeling purposes these are modeled as **hexagons**. **Base stations** are positioned at the **center of these cells**.

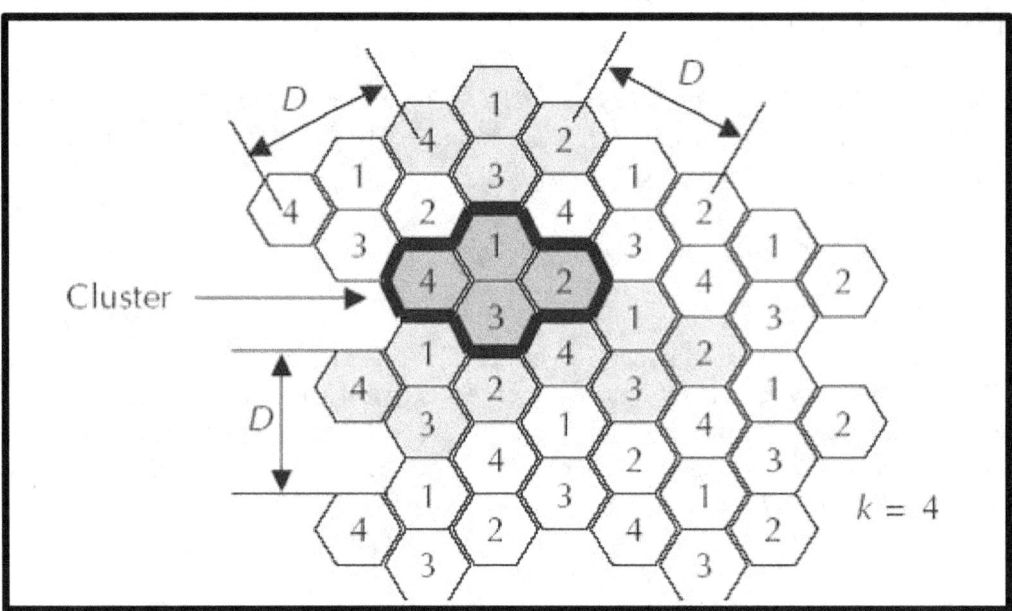

Figure 3.8: Cell Clusters in GSM

Several cells collectively form a cluster (Figure 8 depicts a cluster of size 4). **Each cell i in a cluster receives a subset of total frequency assigned to a mobile network (fbi)**. To avoid any type of co-channel interference, **two neighboring cells never use the same frequencies**. Only at a distance of D (known as **frequency reuse distance**), the same frequency from the set fbi can be reused. **Cells with distance D from cell i, can be assigned the frequencies from the set fbi that belongs to cell i**. **Frequency reuse distance (D)** is given by the number of cells in a cluster.

When moving from one cell to another during an ongoing conversation, an **automatic channel change** occurs. This phenomenon is called **handover**. **Handover** maintains an active speech and data connection over cell boundaries.

3.2.4 Roaming in GSM:

In wireless telecommunications, *roaming* is a general term that refers to the *extending of connectivity service in a location that is different from the user's home location* where the service was registered.

A location that is different from the user's home network is termed as a *Foreign Network*. Roaming ensures that the wireless device keeps connected to the network, without losing the connection even in a foreign network.

Handover refers to the movement from one point of attachment to another point of attachment within a *same mobile network*. *Handover* that occurs *between two different mobile networks* is called roaming.

To ensure security and to deny services to an unauthorized visitor, the *Foreign PLMN (FPLMN)* has to *validate a roamer* as an authorized subscriber *before granting permission* to roam in its network. It is done in the *following manner*:

1. When the mobile device is switched on or is transferred via a handover to the foreign network, this new *"foreign" network* sees the device, notices that it is not registered with its own system, and *attempts to identify its home network*. If there is *no roaming agreement* between the two networks, maintenance of *service is impossible*, and *service is denied* by the visited network.

2. If *there is a roaming agreement* between the two networks, the *foreign network* contacts the *home network* and *requests service information* (including whether or not the mobile should be allowed to roam) about the roaming device *using the IMSI (International Mobile Subscriber Identity) number* obtained from the roaming device.

3. *If successful,* the foreign network begins to *maintain a temporary subscriber record* for the roamer. Likewise, the *home network* updates its information to *indicate that the cell phone is on the foreign network* so that any information sent to that device can be *correctly routed*.

3.2.5 Handover in GSM:

The process of **handover or handoff** within any cellular system is of great importance. It is a critical process and if performed incorrectly handover can result in the loss of the call. Dropped calls are particularly annoying to users and if the number of dropped calls rises, customer dissatisfaction increases and they are likely to change to another network.

Within the GSM system there are four types of handover that can be performed for GSM only systems:

- **Intra-BTS handover**: This form of GSM handover occurs if it is required to **change the frequency or slot being used by a mobile** because of interference, or other reasons. In this form of GSM handover, the **mobile remains attached to the same base transceiver station**, but **change the channel or slot**.

- **Inter-BTS, Intra BSC handover**: This GSM handover or GSM handoff occurs when the **mobile moves out of the coverage area of one BTS** but is **controlled by the same BSC**. In this instance, the BSC is able to perform the handover and it assigns a new channel and slot to the mobile, before releasing the old BTS from communicating with the mobile.

- **Inter-BSC handover**: When the mobile is moved out of the range of cells controlled by one BSC, a more involved form of handover has to be performed, **handing over** not only from one BTS to another but **one BSC to another**. For this the **handover is controlled by the MSC**.

- **Inter-MSC handover**: This form of handover occurs when **changing between networks**. The **two MSCs involved** negotiate to control the handover.

Although there are several forms of GSM handover as detailed above, as far as the mobile is concerned, they are effectively seen as very similar. There are a number of stages involved in undertaking a GSM handover from one cell or base station to another.

In GSM, which uses TDMA techniques the **transmitter only transmits for one slot in eight**, and similarly the **receiver only receives for one slot in eight**. As a result, the Radio Frequency section of the mobile could be **idle for six slots out of the total eight**. This is **not the case** because during the **slots in which it is not communicating** with the BTS, **it scans the other radio channels** looking for frequencies that may be stronger or more suitable.

In addition to this, when the mobile communicates with a particular BTS, it also **sends out a list of the radio channels of the frequencies of neighboring BTSs** via the Broadcast Channel (BCCH). The mobile **scans these and reports back the quality of the links to the BTS**. This way, the **network knows the quality of the link between the mobile and the BTS** as well as the strength of local BTSs as reported back by the mobile.

This is how the **mobile assists in the handover** decision and as a result, this form of GSM handover is known as **Mobile Assisted Hand over (MAHO)**.

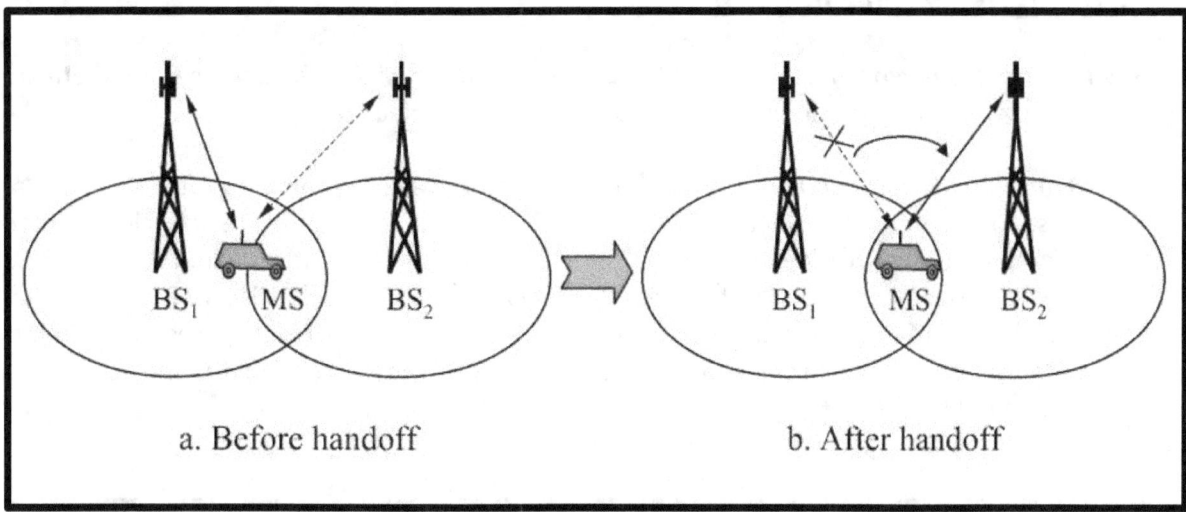

Figure 3.9: Handoff in GSM

3.2.6 Call Routing in GSM:

Human interface is analog while the **information is handled in a digital fashion**. In GSM, there are various complex technologies used between human analog interface & digital network:

- **Digitizer and Source Coding**: The user speech is digitized at 8 KHz sampling rate using Regular Pulse Excited–Linear Predictive Coder (RPE–LPC) with a Long-Term Predictor loop where information from previous samples is used to predict the current sample. Each sample is then represented in signed 13-bit linear PCM value. This digitized data is passed to the coder with frames of 160 samples where encoder compresses these 160 samples into 260-bits GSM frames resulting in one second of speech compressed into 1625 bytes and achieving a rate of 13 Kbits/sec.

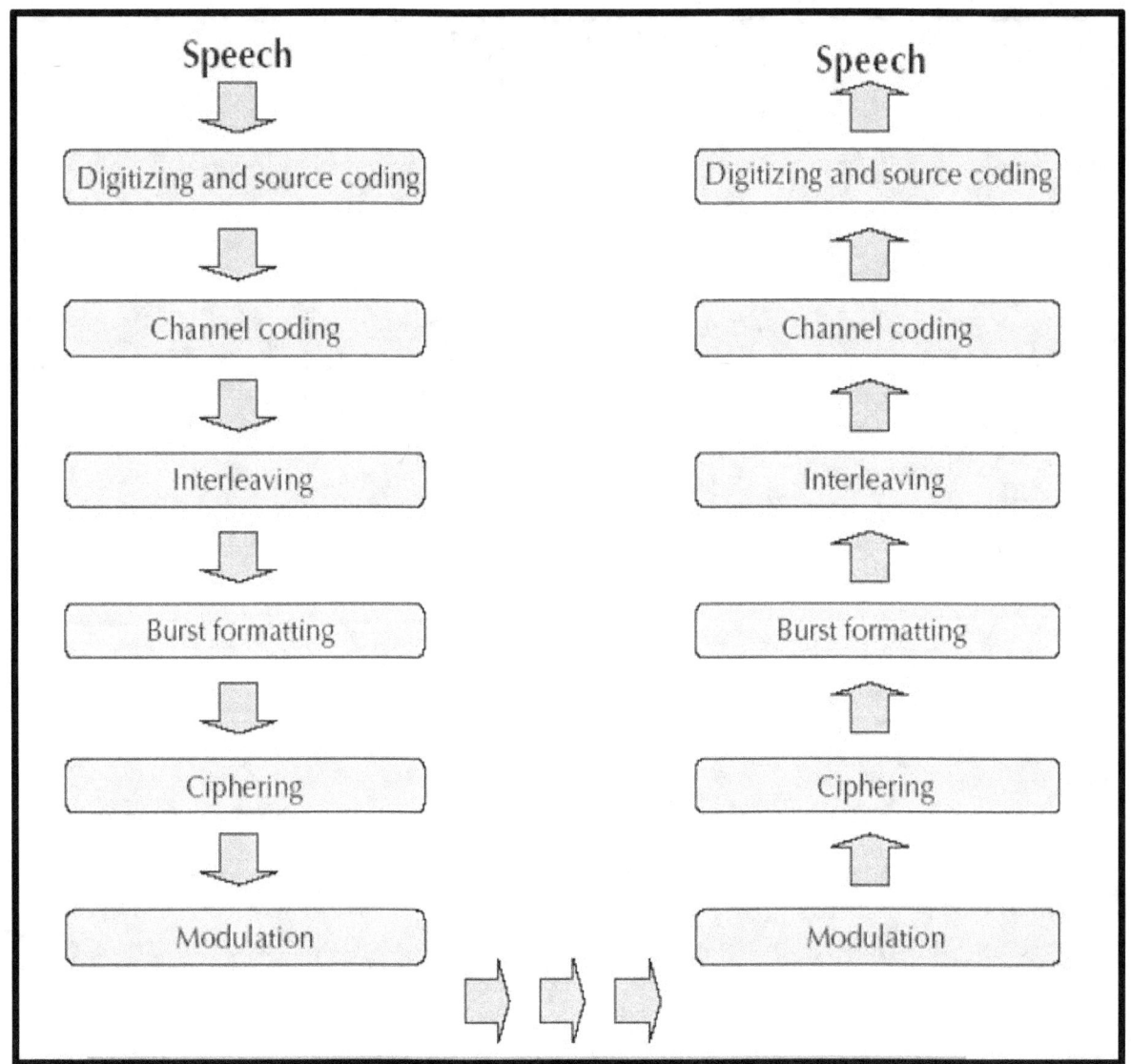

Figure 3.10: Call Routing - Block Diagram

o **Channel Coding**: This introduces **redundancy into the data for error detection** and possible error correction where the gross bit rate after channel coding is 22.8 kbps. These 456 bits are divided into eight 57-bit blocks and the result is interleaved amongst eight successive time slot bursts for protection against burst transmission errors.

o **Interleaving**: This step **rearranges a group of bits in a particular way to improve the performance of the error-correction mechanisms**. The interleaving decreases the possibility of losing whole bursts during the transmission by **dispersing the errors**.

o **Ciphering**: This **encrypts blocks of user data** using a symmetric key shared by the mobile station and the BTS.

o **Burst Formatting**: It **adds some binary information to the ciphered block** for use in synchronization and equalization of the received data.

o **Modulation**: The modulation technique chosen for the GSM system is the **Gaussian Minimum Shift Keying (GMSK)** where **binary data is converted back into analog signal** to fit the frequency and time requirements for the multiple access rules. This signal is then **radiated as radio wave** over the air.

o **Multipath and Equalization**: An **equalizer** is in charge of **extracting the 'right' signal from the received signal** while estimating the channel impulse response of the GSM system and then it constructs an inverse filter. The received signal is then passed through the inverse filter.

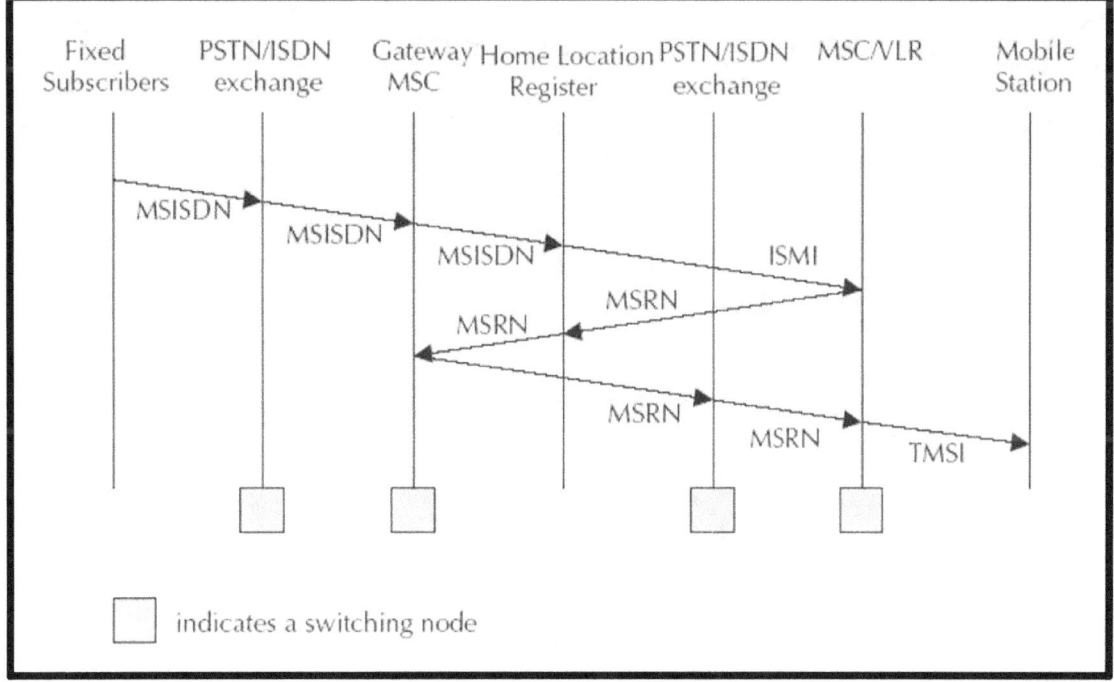

Figure 3.11: Call Routing in GSM

Steps in call routing (Example):

1. The MSISDN number of a subscriber in Bangalore associated with Airtel network is +919845XYYYYY which is a unique number and understood from anywhere in the world.

2. Here, + means prefix for international dialing, 91 is the country code for India and 45 is the network operator's code (Airtel in this case).

3. X is the level number managed by the network operator ranging from 0 to 9 while YYYYY is the subscriber code which, too, is managed by the operator.

4. The call first goes to the local PSTN exchange where PSTN exchange looks at the routing table and determines that it is a call to a mobile network.

5. PSTN forwards the call to the Gateway MSC (GMSC) of the mobile network.

6. MSC enquires the HLR to determine the status of the subscriber. It will decide whether the call is to be routed or not.

7. If MSC finds that the call can be processed, it will find out the address of the VLR where the mobile is expected to be present.

8. If VLR is that of a different PLMN, it will forward the call to the foreign PLMN through the Gateway MSC.

9. If the VLR is in the home network, it will determine the Location Area (LA).

10. Within the LA, it will page and locate the phone and connect the call.

3.2.7 GSM Services:

There are three types of services offered through GSM which are:

1. **Telephony (also referred as tele-services) Services**
2. **Data (also referred as bearer services) Services**
3. **Supplementary Services**

Teleservices or Telephony Services:

A teleservice utilizes the capabilities of a Bearer Service to transport data, defining which capabilities are required and how they should setup.

- **Voice Calls**: The most basic teleservices supported by GSM is telephony. This includes full rate speech at 13 Kbps and emergency calls, where the nearest emergency service provider is notified by dialing three digits.

- **Videotext**: Another group of teleservices includes Videotext access & Teletext transmission.

- **Short Text Messages**: SMS service is a text messaging which allow you to send and receive text messages on your GSM mobile phones.

Data Services:

Using your GSM phone to receive and send data is the essential building block leading to widespread mobile Internet access and mobile data transfer. GSM currently has a data transfer rate of 9.6k. New development that will push up data transfer, HSCSD, are now available.

Supplementary Services:

Supplementary services are provided on top of teleservices or bearer services, and include features such as caller identification, call forwarding, call waiting, multi-party conversation. A brief description of supplementary services is given here:

- **Multiparty Service or conferencing**: The multiparty service allows a mobile subscriber to establish multiparty conservations. That is, conservation between three or more subscribers to setup a conference calls. This service is only applicable to normal telephony.

- **Call Waiting**: This service allows a mobile subscriber to be notified of an incoming call during a conversation. The subscriber can answer, reject or ignore the incoming call. Call waiting is applicable to all GSM telecommunications services using circuit switched connection.

- **Call Hold**: This service allows a mobile subscriber to put an incoming call on hold and then resume this call. The call hold service is only applicable to normal telephony.

- **Call Forwarding**: The call forwarding supplementary service is used to divert calls from the original recipient to another number, and is normally set up by the subscriber himself. It can be used by the subscriber to divert calls from the Mobile Station when the subscriber is not available, and so to ensure that calls are not lost.

- **Call Barring**: The concept of blocking certain type of calls might seem to be a supplementary disservice rather than service. However, there are times when the subscriber is not the actual user of the Mobile Station, and as a consequence may wish to limit its functionality, so as to limit charges incurred. So, GSM devised some flexible services that enable the subscriber to conditionally block calls.

3.2.8 GSM versus CDMA:

FUNCTIONS	GSM	CDMA
FREQUENCY	900MHz; 1800MHz;1900MHz	800MHz;1900MHz
CHANNEL BANDWIDTH	Total 25 MHz bandwidth with 200 KHz per channels, 8 timeslots per channel with frequency hopping.	Total 12MHz with 1.25 MHz for the spread spectrum.
VOICE CODEC	13 Kbps	8 Kbps or 13 Kbps
DATA BIT RATE	9.6 Kbps or expandable	9.6 Kbps
SMS	160 characters of text supports	120 characters
SIM CARD	Yes	No
MULTIPATH	Causes interference and destruction to service	Used as an advantage
RADIO INTERFACE	TDMA	CDMA
HANDOFF	Hard	Soft
SYSTEM CAPACITY	Fixed and limited	Flexible and higher than GSM

3.3 General Packet Radio Service (GPRS):

3.3.1 Introduction to GPRS:

General Packet Radio Service (GPRS) is a means of providing **packet switched data service** with full mobility and wide area coverage on current GSM based wireless networks. GPRS allows for **data speeds of 14.4 Kbps to 171.2 Kbps**, which allow for comfortable Internet access.

Deployment of GPRS networks allows a variety of new applications ranging from **mobile ecommerce** to **mobile corporate VPN access**. It allows for **short burst traffic**, such as **e-mail & web browsing**.

One of the biggest advantage of GPRS is that **no dial-up modem connection** is necessary.

Another prime advantage of GPRS is that it offers **fast connection set-up mechanism** to offer a perception of being *'always on'* or *'always connected'*. Similar to SMS, GPRS is **always 'on service'**.

There are certain **Quality of Service (QoS)** requirements of a mobile packet switched data applications that GPRS needs to follow. GPRS allows definition of QoS profiles using the **following parameters**:

- **Service Precedence**: It is the priority of a service in relation to another service. It can be high, normal or low.

- **Reliability**: It indicates the transmission characteristics required by an application and guarantees certain maximum values for the probability of loss, duplication, dis-sequencing and corruption of packets.

- **Delay**: It define maximum values for the delay. The term delay refers to the end-to-end transfer time between two communicating systems.

- **Throughput**: It specifies the maximum bit rate & the mean bit rate.

3.3.2 GPRS Architecture:

General Packet Radio Service (GPRS) uses the **GSM architecture for voice**.

To offer **packet data services** through GPRS, a **new class of network nodes** called **GPRS support nodes (GSN)** are introduced. GSNs are responsible for the delivery and routing of data packets between the **mobile stations** and the external **packet data networks (PDN)**. Two main GSNs are **Serving GSN (SGSN) & Gateway GSN (GGSN)**.

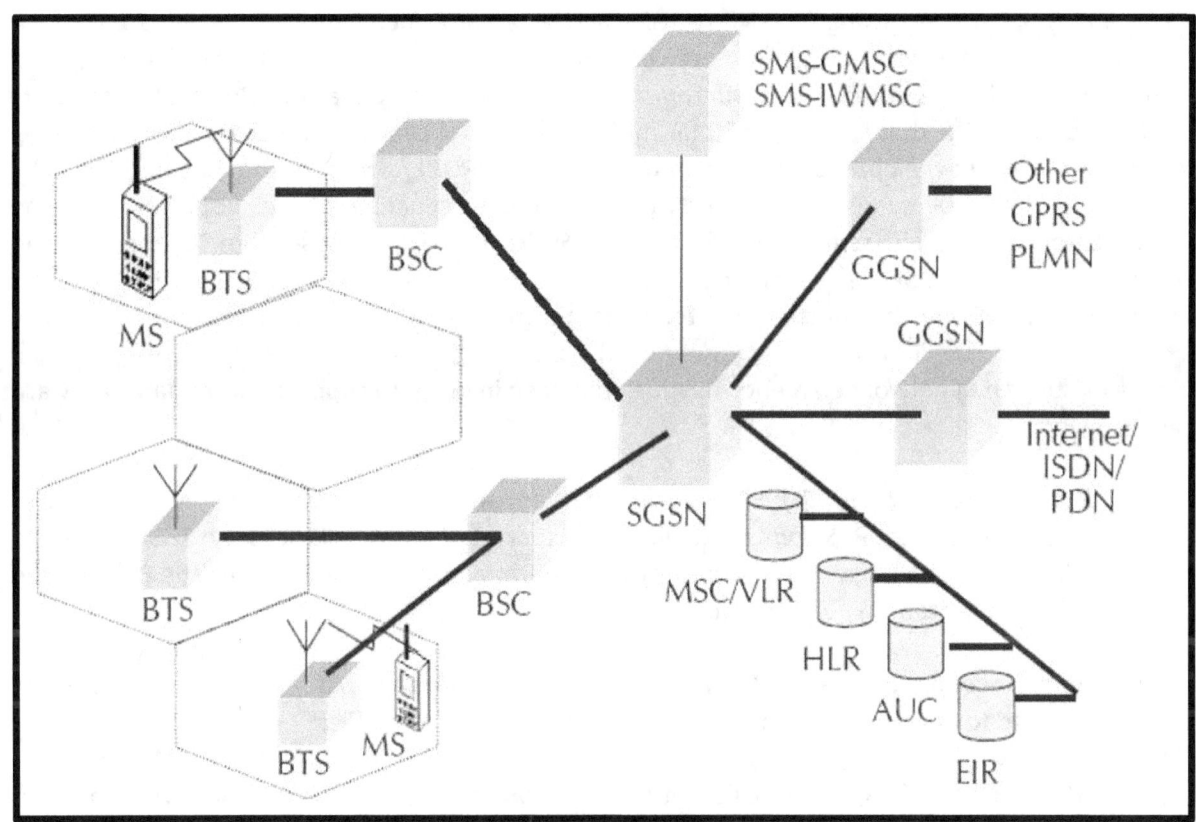

Figure 3.12: GPRS Architecture

Abbreviations:

AUC: Authentication Center
MS: Mobile Station
BSC: Base Station Controller
MSC: Mobile Switching Center
BTS: Base Transceiver Station
PDN: Packet Data Network
EIR: Equipment Identity Register
PLMN: Public Land Mobile Network

GGSN: Gateway GPRS Support Node
SMSC: Short Message Service Center
GPRS: General Packet Radio Service
SMS-GMSC: SMS Gateway MSC
HLR: Home Location Register
SMS-IWMSC: SMS Inter-Working MSC
ISDN: Integrated System Digital Network
SGSN: Serving GPRS Support Node

Serving GPRS Support Node (SGSN): A SGSN is at the ***same hierarchical level as the MSC***. Whatever functions ***MSC does for the voice***, ***SGSN does the same for packet data***. SGSN's tasks include ***packet switching, routing & transfer, mobility management, logical link management***, and ***authentication and charging functions***. SGSN also ***processes registration of new mobile subscribers*** and ***keeps a record*** of their location inside a given service area. The ***location register (LR)*** of the SGSN ***stores location information*** and uses profiles of all GPRS users registered with the SGSN. SGSN sends queries to HLR to obtain profile data of GPRS subscribers. The SGSN is ***connected to the base station*** system with ***Frame Relay***.

Gateway GPRS Support Node (GGSN): A GGSN acts as an ***interface between the GPRS backbone network and the external packet data network***. GGSN's function is ***similar to that of a router in a LAN***. GGSN ***maintains routing information*** that is necessary to tunnel the ***Protocol Data Units (PDUs)*** to the SGSNs that service particular mobile stations. It ***converts the GPRS packets*** coming from the SGSN ***into the appropriate packet data protocol (PDP) format*** for the data networks like internet or X.25, PDP ***sends these packets*** out on the corresponding ***packet data network***. The readdressed packets are ***sent to the responsible SGSN***. For this purpose, the GGSN stores the current SGSN address of the user and his or her profile in its location register. GGSN also performs ***authentication & charging*** functions related to data transfer.

Some existing GSM network elements must be enhanced in order to support packet data. They are as following:

- ***Base Station System (BSS)***: BSS system needs enhancements to recognize and send packet data. This includes ***BTS upgrade*** to allow transportation of ***user data*** to the SGSN. Also, the BTS needs to be upgraded to ***support packet data transportation*** between the BTS and the MS (Mobile Station) ***over the radio***.

- ***Home Location Register (HLR)***: HLR needs enhancement to ***register GPRS user profiles and respond to queries*** originating from GSNs regarding these profiles.

- ***Mobile Station (MS)***: The mobile station or the ***mobile phone for GPRS is different*** from that of GSM.

- ***SMS Nodes***: SMS-GMSCs and SMS-IWMSCs are upgraded to ***support SMS transmission via the SGSN.*** Optionally, the MSC/VLR can be enhanced for more efficient coordination of GPRS and non-GPRS services and functionality. GPRS uses ***two frequency bands*** at 45 MHz apart; viz., ***890-915 MHz for uplink*** (MS to BTS), and ***935-960 MHz for downlink*** (BTS to MS).

3.3.3 PDP Context Activation in GPRS:

In *General Packet Radio Service (GPRS)* network, *MS (Mobile Station)* registers itself with *SGSN (Serving GPRS Support Node)* through a *GPRS attach* which establishes *a logical link between the MS and the SGSN*. To exchange data packets with external *PDNs (Packet Data Networks)* after a successful GPRS attach, an MS must apply for an address which is called *PDP (Packet Data Protocol) address*.

For each session, a *PDP context* is created which contains *PDP type (e.g. IPv4)*, *PDP address assigned to the mobile station (e.g. 129.187.222.10)*, *requested QoS* and *address of the GGSN that will function as an access point to the PDN*.

Such a context is stored in MS, SGSN and GGSN. With an active *PDP (Packet Data Protocol)* context; the MS is visible to the external *PDN (Packet Data Network)*. A user may have *several simultaneous PDP contexts active* at a given time and user data is transferred transparently between MS and external data networks.

Allocation of the PDP address can be *static or dynamic*.

In case of *static address*, the network operator permanently assigns a PDP address to the user.

In case of *dynamic address*, a PDP address is assigned to the user upon the activation of a PDP context request.

Using the message *"activate PDP context request"*, MS informs the SGSN about the requested PDP context and if *request is for dynamic* PDP address assignment, the parameter *PDP address will be left empty*.

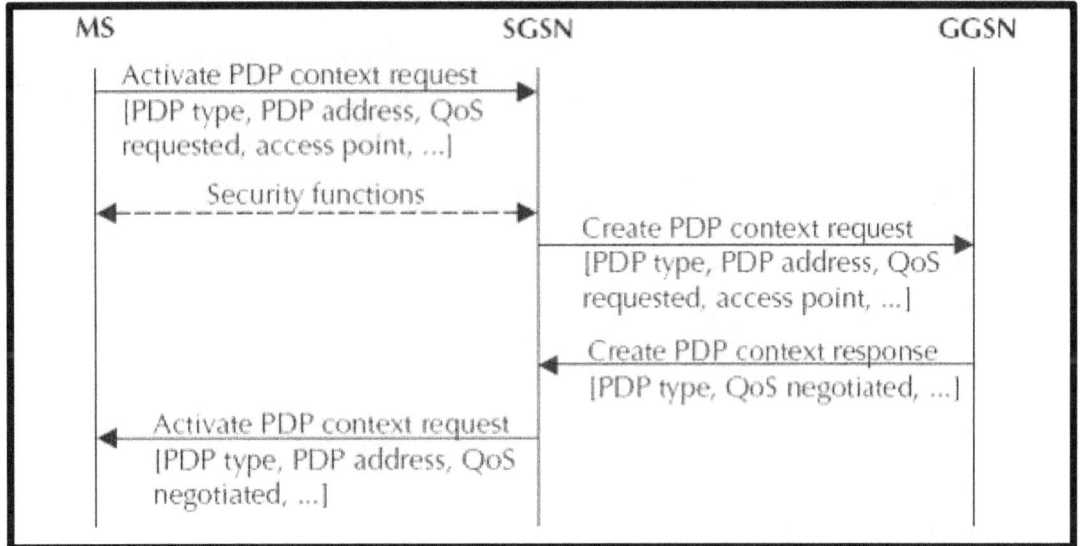

Figure 3.13: PDP Context Activation

After necessary security steps, *if authentication is successful*, SGSN will send a *'create PDP context request'* message to the GGSN, the result of which is a *confirmation message 'create PDP context response'* from the GGSN to the SGSN, which contains the *PDP address*.

SGSN updates its PDP context table and *confirms the activation* of the new PDP context *to the MS*.

Disconnection from the GPRS network is called *GPRS detach* in which all the resources are released.

3.3.4 Routing in GPRS:

Routing is the process of how packets are routed in GPRS. Here, the example assumes two *intra-PLMN backbone* networks of different PLMNs. *Intra-PLMN backbone* networks *connect GSNs (GPRS Service Nodes) of the same PLMN (Public Land Mobile Network)*.

These *intra-PLMN networks* are connected with an *inter-PLMN backbone* which *connects GSNs of different PLMNs*. However, a *roaming agreement* is necessary between two GPRS network providers.

Gateways between PLMNs and external inter-PLMN backbone are called *border gateways* which perform *security functions* to protect the private intra-PLMN backbones *against malicious attacks*.

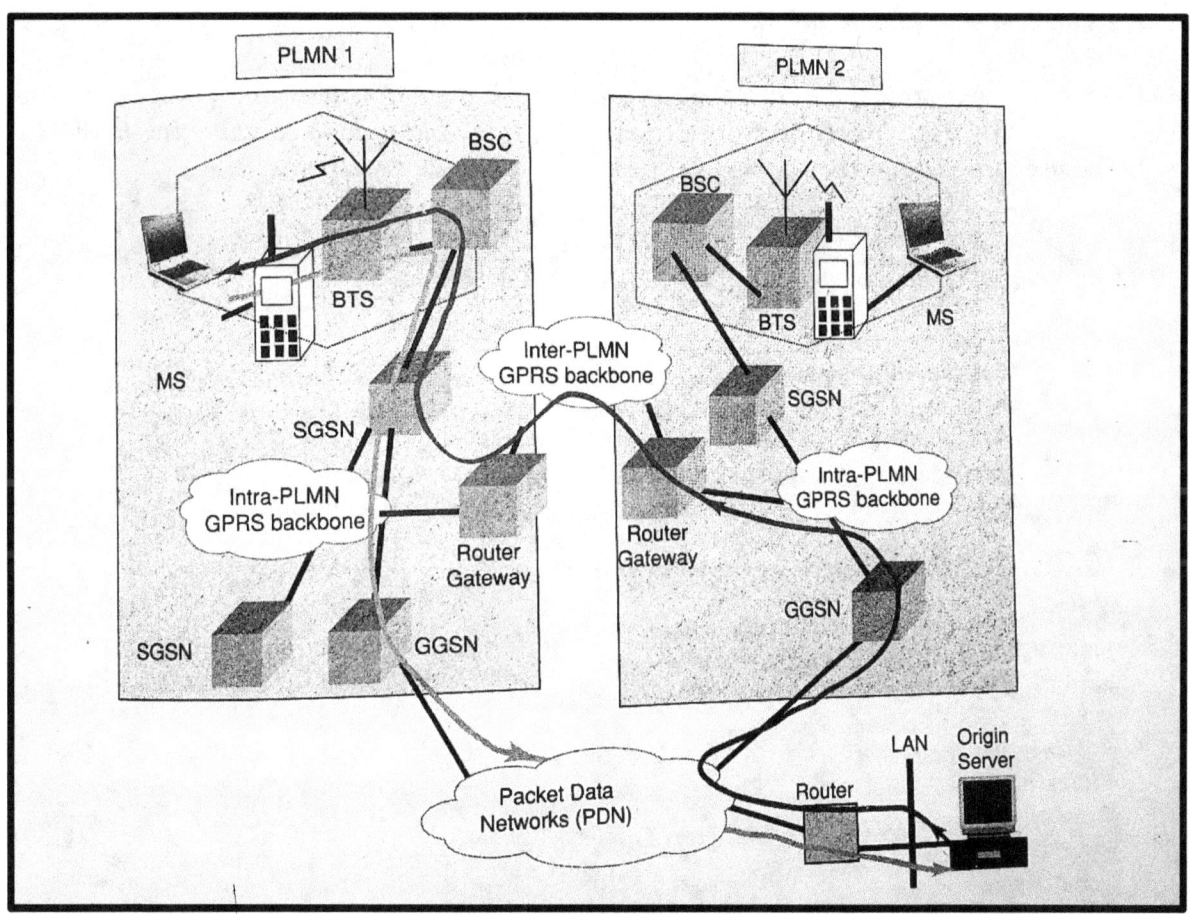

Figure 3.14: Routing in GPRS

Let's say that *MS (Mobile Station)*, that is *currently located in PLMN1 sends IP packets* to a host connected to the IP network (e.g. to a Web server connected to the Internet). We assume that the Packet Data Network (PDN) is an IP network.

The **SGSN**, that the MS is registered with currently, **encapsulates the IP packets** coming from the mobile station and **routes them through the intra-PLMN GPRS backbone to the appropriate GGSN**. **GGSN de-encapsulates the packets and sends them out on the IP network**, where IP routing mechanisms are used to **transfer the packets to the access router of the destination network** and finally, **delivers the IP packets to the host**.

Let us also say that **home-PLMN of the mobile station** (that is currently in PLMN 1) **is PLMN2**. **Correspondent host is now sending IP packets to the MS** onto the IP network and are **routed to the GGSN of PLMN2** (the home-GGSN of the MS). **But the MS is currently in PLMN 1**.

In such case, **GGSN of PLMN2 queries the HLR and obtains the information that the MS is currently located in PLMN1**. It encapsulates the **incoming IP packets** and tunnels them through the **inter-PLMN GPRS backbone** to the **appropriate SGSN in PLMN1** while the **SGSN de-encapsulates** the packets and **delivers them to the MS**.

HLR stores the **user profile**, the **current SGSN address** and the **PDP addresses** for every GPRS user in the PLMN. When the **MS registers with a new SGSN (eg. while visiting another PLMN)**, HLR will send the **user profile** to the **new SGSN**.

3.3.5 Data Services in GPRS:

Any user is likely to use either of the **two modes** of the GPRS network:

- o **Application mode**
- o **Tunneling mode**

In **application mode**, user uses the GPRS mobile phone to access the applications running on the phone itself. The **phone here acts as the end user device**.

In **tunneling mode**, user uses GPRS interface as an access to the network as the **end user device** would be a **large footprint device** like **laptop computer** or a **small footprint device** like **PDA**. The mobile phone will be connected to the device and used as a modem to access the wireless data network.

3.3.6 Billing & Charging in GPRS:

For voice networks tariffs are generally based on distance and time means that user pay more for long distance calls. In data services, minimum charging information that must be collected are:

- Destination and source addresses
- Usage of radio interface
- Usage of external Packet Data Networks (PDNs)
- Usage of the packet data protocol addresses
- Usage of general GPRS resources
- Location of the Mobile Station

A GPRS network needs to be able to count packets to charge customers for the volume of packets they send and receive.

Various business models exist for charging customers as billing of services can be based on the **transmitted data volume**, the **type of service**, the **chosen QoS profile**, etc.

GPRS call records are generated in the **GPRS Service Nodes**. **Packet counts** are passed to a **Charging Gateway** that **generates Call Detail Records** that are **sent to the billing system**.

3.3.7 Applications & Limitations of GPRS:

Applications of GPRS:

I. **Communications**: E-mail, fax, unified messaging and intranet/internet access, etc.

II. **Value-added services**: Information services and games, etc.

III. **E-commerce**: Retail, ticket purchasing, banking and financial trading, etc.

IV. **Location-based applications**: Navigation, traffic conditions, airline/rail schedules and location finder, etc.

V. **Vertical applications**: Freight delivery, fleet management and sales-force automation.

VI. **Advertising**: It may be location sensitive. For example, a user entering a mall can receive advertisements specific to the stores in that mall.

VII. **Chat**: It is used as means to communicate and discuss matters of common interest. GPRS will offer ubiquitous chat by integrating Internet chat and wireless chat using SMS and WAP.

VIII. **Multimedia Services**: Multimedia objects like photographs, pictures, postcards, greeting cards, presentations & static web pages can be sent and received over the mobile network.

IX. **Virtual Private Network**: GPRS is used to provide VPN services. As the bandwidth is higher so many banks in India are migrating to GPRS-based VPNs. This is expected to reduce the transaction time by about 25%.

X. **Vehicle Positioning**: This application integrates GPS that tell people where they are. Anyone with a GPS receiver can receive their satellite position and thereby find out where they are. Vehicle Positioning applications can be used to deliver several services including remote vehicle diagnostics.

Limitations of GPRS:

I. **Limited Cell Capacity for All Users:** GPRS does impact a network's existing cell capacity. There are only limited radio resources that can be deployed - use for one purpose (data services) precludes simultaneous use for another (voice). Voice and GPRS calls both use the same network resources. If tariffing and billing are not done properly, this may have impact on revenue.

II. **Speeds Much Lower in Reality**: Achieving the theoretical maximum GPRS data transmission speed of 172.2 kbps would require a single user taking over all eight timeslots without any error protection. Clearly, it is unlikely that a network operator will allow all timeslots to be used by a single GPRS user. Additionally, the initial GPRS terminals are expected be severely limited - supporting only one, two or three timeslots. The bandwidth available to a GPRS user will therefore be severely limited. The reality is that mobile networks are always likely to have lower data transmission speeds than fixed networks.

III. **_Transit Delays_**: GPRS packets are sent in all different directions to reach the same destination. This opens up the potential for one or some of those packets to be lost or corrupted during the data transmission over the radio link. The GPRS standards recognize this inherent feature of wireless packet technologies and incorporate data integrity and retransmission strategies. However, the result is that potential transit delays can occur.

3.4 Mobile IP:

3.4.1 Working of Mobile IP (Architecture):

What is the problem?
You are connected to a Wi-Fi with a certain IP address. You then move to a different IP address. You still want to maintain your "reachability". But TCP connections are bound to your IP address and break when your IP changes. You cannot keep your old IP in a new network, as IP address allocation is hierarchical. The root of the problem is that your IP address serves as both your "address" that is used to locate you, and as your "identity" that is used to identify you in TCP connections.

Mobile IP signifies that, while a user is connected to applications across the Internet and the user's point of attachment changes dynamically, all connections are maintained despite the change in underlying network properties.

To solve the above stated problem, Mobile IP allows the mobile node to use two IP addresses called **home address** and **care-of address**.

The **_home address_** is static IP address and is known to everybody as the identity of the host.

Care-of address (COA) of a node is determined by the location of that node & **changes with every new point of attachment**. This address is provided by the Foreign Agent when the mobile node (MN) enters a Foreign Network.

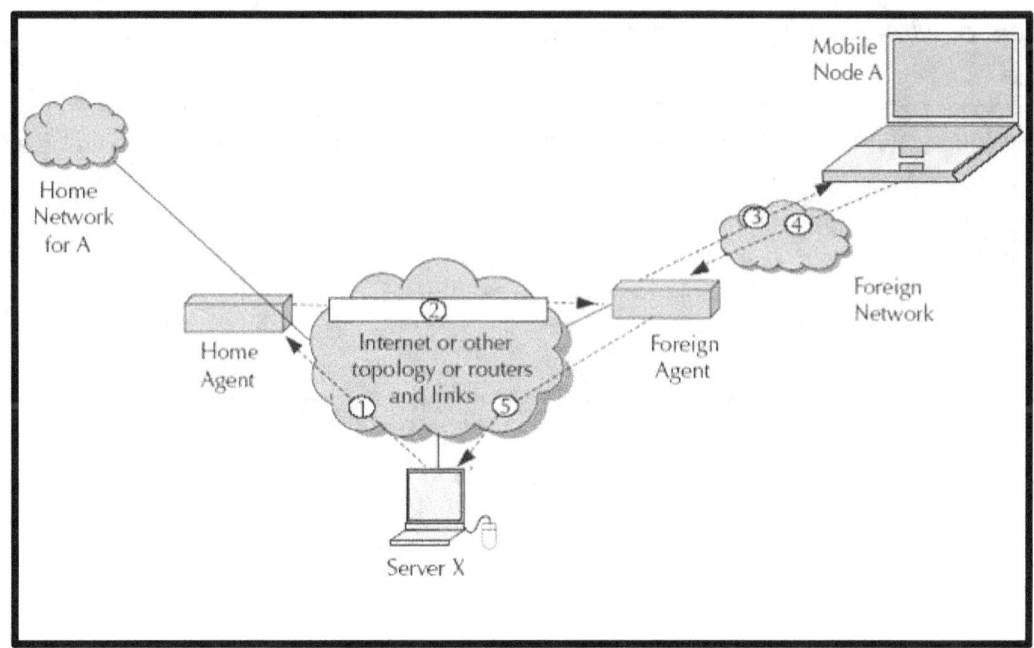

Figure 3.15: Architecture of Mobile IP

There are two mobility agents used by Mobile IP. They are as follows:

- ***Home agent (HA)*** is an entity located on home network of the user. It forwards data packets to appropriate network when mobile node (MN) is away from home network. It communicates with mobile node (MN) all the time regardless of its location.

- ***Foreign agent (FA)*** is the counterpart of the Home Agent (HA) in the foreign network. When a mobile node (MN) enters foreign network, it registers to foreign agent. FA provides care-of address (COA) to the MN & also communicates this address to home agent.

Steps in Mobile IP operation:

1. When a mobile node moves to a new care-of address (COA), it informs the home agent of its new address and period of validity. The home agent maintains the binding (home address, care-of address (COA), duration of validity).

2. Note that the mobile host cannot announce a route from its home address to the new network (that would be done by the Home Agent), so packets destined to its home address won't automatically reach the user. When any packet arrives at the home address, the home agent tunnels that packet to the mobile node at its new address via the foreign agent.

3. The mobile node can send replies directly, or reverse tunnel them via home agent.

4. IP-in-IP tunneling means that the original IP packet is encapsulated in another IP packet with the new destination address being the care of address.

Example of Mobile IP (Consider Figure 15):

1. Server X wants to transmit packet to node A. Home address of A is known to X but it does not know whether A is in home network or not. Thus, X sends packet to A's home address as destination address.

2. Packet is intercepted by home agent which knows that A is in foreign network. Care-off address (COA) of A is available to the home agent (HA). Packet is retransmitted to node A by the home agent (HA).

3. In foreign network, the packet is intercepted by foreign agent (FA) & is delivered to node A.

4. Node A responds to this message to server X via foreign agent (FA).

5. Foreign agent (FA) sends reply to server X.

3.4.2 Discovery, Registration & Tunneling:

Three operations performed by the Mobile IP are – *Discovery, Registration & Tunneling*.

A. Discovery:

- A mobile node uses a discovery procedure to identify prospective home agents and foreign agents.

- Router/Agent advertisements are transmitted by both HA & FA to advertise their services on a link. Mobile nodes use these advertisements to determine their current point of attachment on the network.

- This helps the mobile node to determine whether it is in a foreign network or the home network.

- On receiving the advertisement ICMP message from the agent, mobile node compares the network portion of the agent IP address to the network portion of its own IP address provided by the home network. If these network portions do not match, the mobile node knows that it is in a foreign network.

- A router/agent advertisement can carry information about one or more care-of addresses. If a mobile node needs a care off address without waiting for router/agent advertisement, then it can broadcast a solicitation that will be answered by foreign agent.

B. Registration:

- When mobile node (MN) obtains care-of address (COA), it needs to be registered at the home agent (HA).

- The mobile node sends a registration request to its home agent with the care-of address information. This information contains a triplet – (home address, care-of address, registration validity).

- On receiving this registration request, the HA updates its routing table (this is called binding for the mobile node) & sends a registration reply back to the mobile node. The binding data is maintained by the HA until the registration validity expires.

- As a part of registration process, the mobile node needs to be authenticated. This is done using a digital signature generated by a 128-bit secret key & HMAC-MD5 hashing algorithm.

- Each home agent & mobile node shares a common secret which makes the digital signature unique.

C. *Tunneling:*

- Tunneling procedure is used to forward IP datagrams from a home address of a mobile node to its care-of address.

- Packets destined to a node's home address won't automatically reach the mobile node. When any packets arrive at the home address, the home agent tunnels those packets to the mobile node at its new address via the foreign agent. This is known as Tunneling.

- IP-in-IP tunneling is used here. This means that the original IP packet is encapsulated in another IP packet with the new destination address being the care of address.

3.4.3 Traditional IP versus Mobile IP:

Traditional IP	Mobile IP
Here, Routing mechanisms rely on the assumption that each network node will always have the same point of attachment to the Internet, and each node's IP address identifies the network link where it is connected. Core Internet routers look at the IP address prefix, which identifies a device's network. At the network level, routers look at the next few bits to identify the appropriate subnet. Finally, at the subnet level, routers look at the bits identifying a particular device. In this routing scheme, if you disconnect a mobile device from the Internet and want to reconnect through a different subnet, you have to configure the device with a new IP address, the appropriate netmask and default router. Otherwise, routing protocols have no means of delivering packets because the device's IP address doesn't contain the necessary information about the current point of attachment to the Internet.	Mobile IP assign each mobile node with a permanent home address on its home network and a care-of address that identifies the current location of the device within a network and its subnets. Each time a user moves the device to a different network, it acquires a new care-of address. A mobility agent on the home network associates each permanent address with its care-of address. The mobile node sends the home agent a binding update each time it changes its care-of address using Internet Control Message Protocol (ICMP). In Mobile IPv4, traffic for the mobile node is sent to the home network but is intercepted by the home agent and forwarded via tunneling mechanisms to the appropriate care-of address. This is how the problem of traditional IP is resolved in Mobile IP.

Introduction to Wifi

4.1 Wireless LAN (Wi-Fi) - Advantages, Disadvantages & Goals:

Wireless LAN is also referred to as **Wireless Fidelity (Wi-Fi)**.

Wi-Fi is a local area data network **without any physical connectivity**, i.e. wires. **WLAN** is implemented as an **extension to a wired LAN** within a building or campus.

WLANs are typically **restricted in their diameter**: building, campus, single room etc.

The global goal is to replace office cabling and to introduce high flexibility for **ad hoc communication** (no centralized server, peer-to-peer network set up **temporarily**).

Wireless LAN Advantages:

- **Mobility**: WLAN offers wire-free access within operating range.
- **Low Implementation Costs**: Eliminates the cost for wired connections.
- **Easy to setup, relocate & change**.
- **Installation Speed and Simplicity**: Fast and simple installation of WLAN.
- **Network Expansion**: Easy expansion of WLAN possible.
- **Higher Flexibility**: within radio coverage, nodes can communicate without further restriction. Radio waves can penetrate walls.
- **Planning**: wireless ad hoc networks allow for communication without planning. Wired networks need wiring plans.
- **Robustness**: wireless networks can survive disasters; if the wireless devices survive people can still communicate.

Wireless LAN disadvantages:

- **QoS**: WLANs offer typically lower QoS. **Lower bandwidth** due to limitations in radio transmission (1- 10 Mbps). **Higher error rates** due to interference.
- **Cost**: Wireless LAN adapters are expensive than Ethernet adapters.

- **Proprietary solutions**: Slow standardization procedures lead to many proprietary solutions that works only in a homogeneous environment.
- **Safety & Security**: Using radio waves for data transmission might interfere with other high-tech equipment.

Wireless LAN: Design Goals:

- **Global Operation**: LAN equipment may be carried from one country to another and this operation should be legal (consider national and international frequency regulations).

- **Low Power**: Take into account that the devices communicating via WLAN are typically running on battery power. Special power saving modes and power management functions should be implemented.

- **Simplified Spontaneous Co-operation**: No complicated setup routines but operate spontaneously after power-on.

- **Easy to Use**: WLANs are made for simple users; they should not require complex management but rather work on a plug-and-play basis.

- **Protection of Current Investment**: A lot of money has been invested for wired LANs; WLANs should be able to interoperate with existing network (same data type and services).

- **Safety & Security**: Safe to operate. Encryption mechanism is as such that it does not allow roaming profiles for tracking people, ensuring privacy.

- **Transparency for Applications**: Existing applications should continue to work.

4.2 Types of Wireless LANs – Adhoc Mode versus Infrastructure Mode:

Types of Wireless LAN are:

- 802.11
- HyperLAN
- HomeRF
- Bluetooth
- MANET

Adhoc Mode versus Infrastructure Mode:

In **Adhoc Mode**, there is **no access point** or infrastructure. A number of mobile stations from a cluster **communicate with each other separately**.

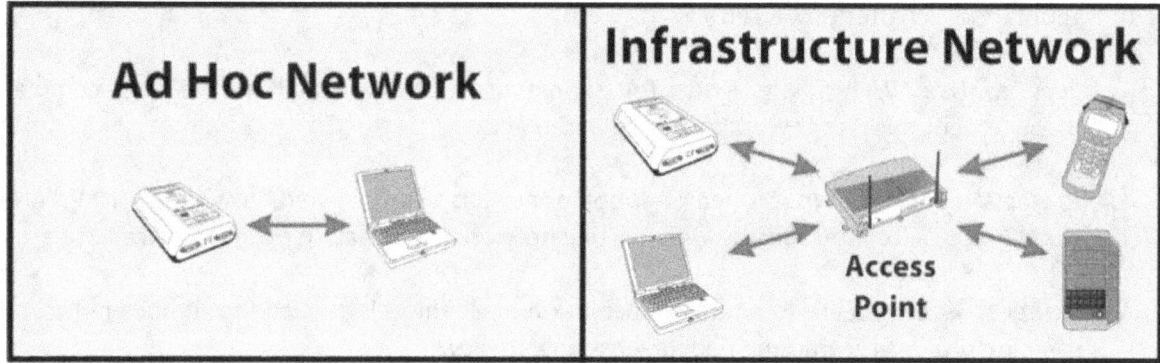

Figure 4.1: Adhoc Mode versus Infrastructure Mode

In **Infrastructure Mode**, the **mobile stations (MSs)** are connected to a **base station or access point**. This is similar to a star network where all the mobile stations are attached to the base station. Through a protocol, the base station manages the **dialogue between the Access Points (AP) and Mobile Stations (MS)**.

4.3 IEEE 802 Architecture:

The architecture of a LAN is best described in terms of a *layering of protocols* that organize the basic functions of a LAN. Such architecture was developed by the *IEEE 802 committee* and has been adopted by all organizations working on the specification of LAN standards. It is generally referred to as the *IEEE 802 reference model*.

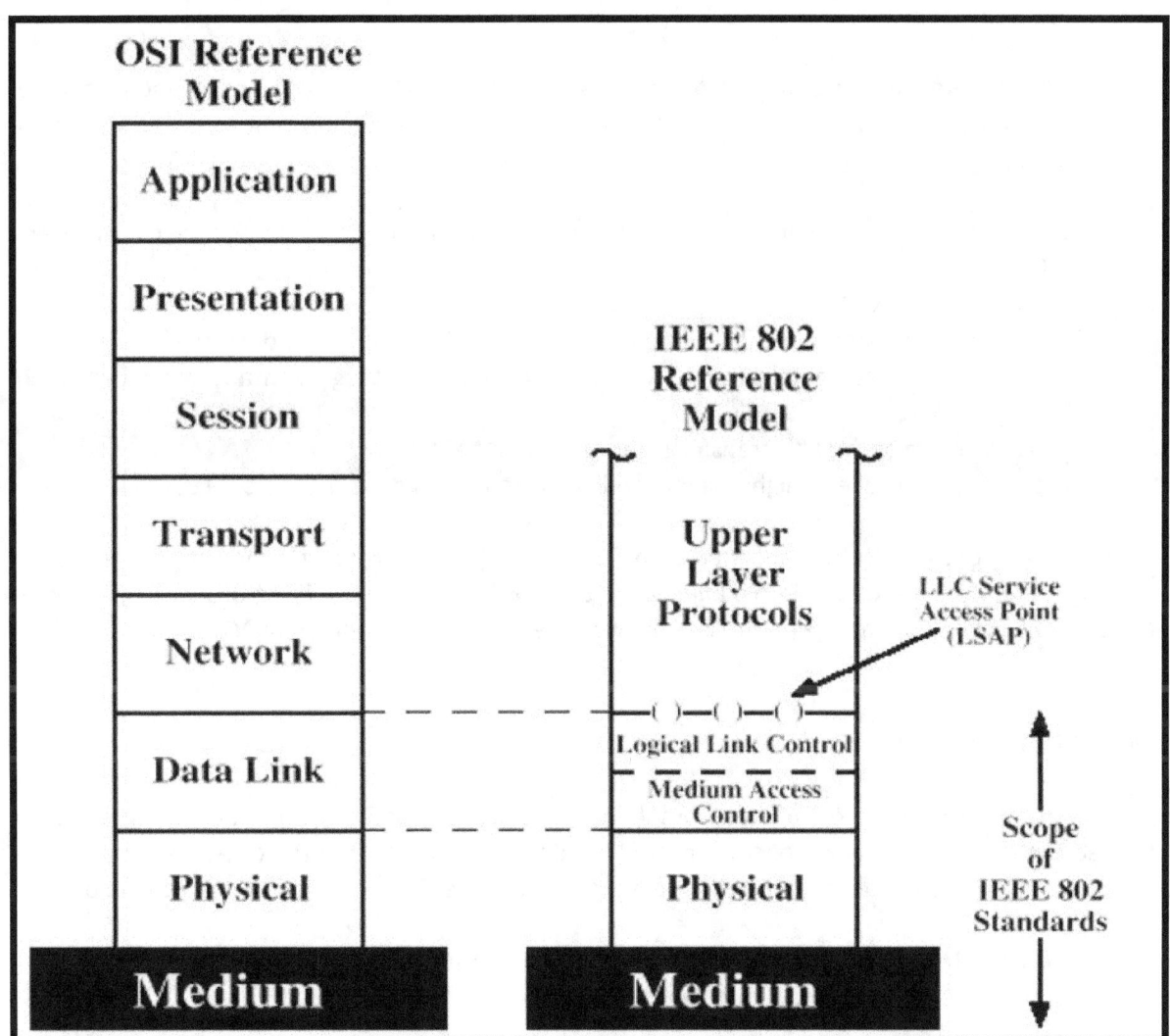

Figure 4.2: IEEE 802 Architecture in Reference to OSI model

Working from the bottom up, the lowest layer of the IEEE 802 reference model corresponds to the *physical layer* of the OSI model and includes functions such as:

- *Encoding/decoding of signals*
- *Preamble (synchronization signal in header) generation/removal (for synchronization)*
- *Bit transmission/reception*

For some of the IEEE 802 standards, the physical layer is further subdivided into sub layers. In the case of IEEE 802.11, two sub layers are defined:

- ***Physical layer convergence procedure (PLCP)***: Defines a method of mapping 802.11 ***MAC layer protocol data units (MPDUs)*** into a framing format suitable for sending and receiving user data and management information between two or more stations using the associated PMD sub layer.

- ***Physical medium dependent sub layer (PMD)***: Defines the characteristics of, and method of transmitting and receiving, user data through a wireless medium between two or more stations.

On top of the physical layer, there are the functions associated with providing service to LAN users. These include:

- On transmission, assemble data into a frame with address and error detection fields. (MAC)
- On reception, disassemble frame, and perform address recognition and error detection. (MAC)
- Govern access to the LAN transmission medium. (MAC)
- Provide an interface to higher layers and perform flow and error control. (LLC)

These are functions typically associated with OSI layer 2 (Data Link Layer). The set of functions in the *last bullet item* is grouped into a *logical link control (LLC) layer*. The functions in the *first three bullet items* are treated as a separate layer, called *medium access control (MAC)*.

MAC Frame Format:

The MAC layer receives a block of data from the LLC layer and is responsible for performing functions related to medium access and for transmitting the data. On reception of the data, it assembles the data into a frame with address and error detection fields. The fields of this frame are as follows:

- ***MAC Control***: This field contains any protocol control information needed for the functioning of the MAC protocol. For example, a priority level could be indicated here.

- ***Destination MAC Address***: The destination physical attachment point on the LAN for this frame.

- ***Source MAC Address***: The source physical attachment point on the LAN for this frame.

- ***Data***: The body of the MAC frame. This may be LLC data from the next higher layer or control information relevant to the operation of the MAC protocol.

- **CRC**: The cyclic redundancy check field (also known as the frame check sequence, FCS, field). This is an error-detecting code.

Logical Link Control (LLC):

Like all other link layers, LLC is concerned with the transmission of a link-level PDU (Protocol Data Unit) between two stations, without the necessity of an intermediate switching node. LLC has two characteristics not shared by most other link control protocols:

1. It must support the multi-access, shared-medium nature of the link (this differs from a multidrop line in that there is no primary node).

2. It is relieved of some details of link access by the MAC layer.

Addressing in LLC involves specifying the source and destination LLC users. Typically, a user is a higher-layer protocol or a network management function in the station. These LLC user addresses are referred to as service access points (SAPs).

LLC specifies the mechanisms for addressing stations across the medium and for controlling the exchange of data between two users. The operation and format of this standard is based on HDLC (High Level Data Link Control).

4.4 IEEE 802.11 Architecture & Services:

IEEE 802.11 Terminology:

Access Point (AP): Any entity that has station functionality and provides access to the distribution system via the wireless medium for associated stations.

Basic Service Set (BSS): A set of stations controlled by a single coordination function.

Coordination Function: The logical function that determines when a station operating within a BSS is permitted to transmit and may be able to receive PDUs.

Distribution System (DS): A system used to interconnect a set of BSSs and integrated LANs to create an ESS.

Extended Service Set (ESS): A set of one or more interconnected BSSs and integrated LANs that appear as a single BSS to the LLC layer at any station associated with one of these BSSs.

MAC Protocol Data Unit (MPDU): The unit of data exchanged between two peer MAC entities using the services of the physical layer.

MAC Service Data Unit (MSDU): Information that is delivered as a unit between MAC users.

Station: Any device that contains an IEEE conformant MAC and physical layer.

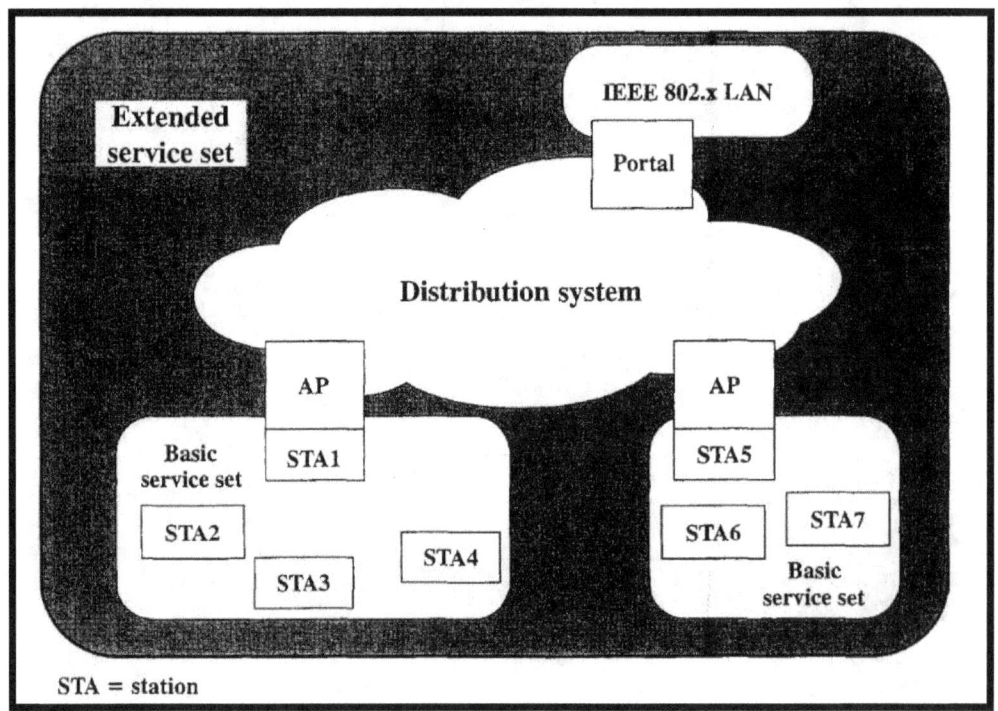

Figure 4.3: IEEE 802.11 Architecture

IEEE 802.11 Architecture:

The smallest building block of a wireless LAN is a *basic service set (BSS)*, which consists of some number of *stations executing the same MAC protocol* and competing for access to the same shared wireless medium.

A BSS may be isolated or it may connect to a *distribution system (DS)* through an *access point (AP)*.

The *distribution system (DS)* is typically a wired backbone LAN.

The *access point (AP)* functions as a *bridge* and a *relay point*. In a BSS, client stations do not communicate directly with one another. Rather, if one station in the BSS wants to communicate with another station in the same BSS, the MAC frame is first sent from the originating station to the AP, and then from the AP to the destination station. Similarly, a MAC frame from a station in the BSS to a remote station is sent from the local station to the AP and then relayed by the AP over the DS on its way to the destination station. AP also provides DS services in addition to acting as a station.

When all the stations in the BSS are *mobile stations*, with *no connection to other BSSs*, the BSS is called an *independent BSS (IBSS)*. An *IBSS* is typically an *ad hoc network* in which all the stations communicate directly, and *no AP is involved*.

It is also possible for *two BSSs to overlap geographically*, so that a single station could participate in more than one BSS.

An *extended service set (ESS)* consists of *two or more BSSs interconnected* by a distribution system. ESS appears as a *single logical LAN* to the *logical link control (LLC)* level.

IEEE 802.11 Services:

Association: Establishes an initial association between a station and an AP. Before a station can transmit or receive frames on a wireless LAN, its identity and address must be known.

Authentication: In a wired LAN, it is generally assumed that access to a physical connection conveys authority to connect to the LAN. This is not a valid assumption for a wireless LAN, in which connectivity is achieved simply by having an attached antenna that is properly tuned. The authentication service is used by stations to establish their identity with stations they wish to communicate with. Authentication scheme could range from relatively unsecure handshaking to public key encryption schemes.

Deauthentication: This service is invoked whenever an existing authentication is to be terminated.

Service	Provider	Used to support
Association	Distribution System	MSDU delivery
Authentication	Station/AP	LAN access and security
Deauthentication	Station/AP	LAN access and security
Disassociation	Distribution System	MSDU delivery
Distribution	Distribution System	MSDU delivery
Integration	Distribution System	MSDU delivery
MSDU delivery	Station/AP	MSDU delivery
Privacy	Station/AP	LAN access and security
Reassociation	Distribution System	MSDU delivery

MSDU – MAC Service Data Unit

Figure 4.4: IEEE 802.11 Services

Disassociation: A notification from either a station or an AP that an existing association is terminated. A station should give this notification before leaving an ESS or shutting down.

Distribution: Distribution is the primary service used by stations to exchange MAC frames when the frame must traverse the DS to get from a station in one BSS to a station in another BSS.

MSDU (MAC Service Data Unit) Delivery: For example, suppose a frame is to be sent from station 2 (STA 2) to STA 7 in Figure 3. The frame is sent from STA 2 to STA 1, which is the AP for this BSS. The AP gives the frame to the DS, which has the job of directing the frame to the AP associated with STA 5 in the target BSS. STA 5 receives the frame and forwards it to STA 7.

Integration: The integration service enables transfer of data between a station on an IEEE 802.11 LAN and a station on an integrated IEEE 802.x LAN. The term integrated refers to a wired LAN that is physically connected to the DS and whose stations may be logically connected to an IEEE 802.11 LAN via the integration service. The integration service takes care of any address translation and media conversion logic required for the exchange of data.

Privacy: Used to prevent the contents of messages from being read by stations other than the intended recipient. The standard provides for the optional use of encryption to assure privacy.

Reassociation: Enables an established association to be transferred from one AP to another, allowing a mobile station to move from one BSS to another.

4.5 IEEE 802.11 Medium Access Control:

The *IEEE 802.11 MAC layer* covers three functional areas: *reliable data delivery*, *medium access control*, and *security*.

Reliable Data Delivery:

As with any wireless network, a wireless LAN using the IEEE 802.11 physical and MAC layers is subject to considerable unreliability. Noise, interference, and other propagation effects result in the loss of a significant number of frames. Even with error-correction codes, a number of MAC frames may not successfully be received. This situation can be dealt with by reliability mechanisms at a higher layer, such as TCP.

To deal with errors at the MAC level, IEEE 802.11 includes a *frame exchange protocol*. When a station receives a data frame from another station, it returns an *acknowledgment (ACK) frame* to the source station. If the source does not receive an ACK within a short period of time, either because its data frame was damaged or because the returning ACK was damaged, the source retransmits the frame. Thus, the basic data transfer mechanism in IEEE 802.11 involves an *exchange of two frames*.

To further enhance reliability, *a four-frame exchange* may be used. In this scheme, a source first issues a request to send (RTS) frame to the destination. ▫ The destination then responds with a clear to send (CTS). After receiving the CTS, the source transmits the data frame, and the destination responds with an ACK.

Medium Access Control:

The 802.11 working group considered two types of proposals for a MAC algorithm:

- *Distributed access protocols*, which distribute the *decision to transmit* over all the nodes using a carrier-sense mechanism.

- *Centralized access protocols*, which involve regulation of transmission by a centralized decision maker.

A *distributed access protocol* makes sense for an ad hoc network of peer workstations (typically an IBSS) and may also be attractive in other wireless LAN configurations that consist primarily of busty traffic.

A *centralized access protocol* is natural for configurations in which a number of wireless stations are interconnected with each other and some sort of base station that attaches to a backbone wired LAN; it is especially useful if some of the data is time sensitive or high priority.

4.6 Wi-Fi Protected Access (WPA):

The original 802.11 specification included a set of security features for privacy and authentication which were quite weak. For privacy 802.11 defined the **Wired Equivalent Privacy (WEP)** algorithm. WEP makes use of the **RC4 encryption algorithm** using a **40-bit key**. A later revision enables the use of a **104-bit key**.

For authentication, 802.11 requires that the two parties share a secret key and defines a protocol by which this key can be used for **mutual authentication**. The privacy portion of the 802.11 standard contained major weaknesses. The 40-bit key is resentfully (badly) inadequate. Even the 104-bit key proved to be vulnerable, due to a variety of weaknesses.

The 802.11i task group has developed a set of capabilities to address the WLAN security issues. In order to accelerate the introduction of **strong security into WLANs**, the **Wi-Fi Alliance** promulgated **Wi-Fi Protected Access (WPA)** as a **Wi-Fi standard**.

WPA is a set of security mechanisms that eliminates most 802.11 security issues and was based on the current state of the 802.11i standard. As 802.11i evolves, WPA will evolve to maintain compatibility. IEEE 802.11i addresses three main security areas: **authentication**, **key management**, and **data transfer privacy**. To improve authentication, 802.11i requires the use of an **authentication server (AS)** and defines a more robust authentication protocol.

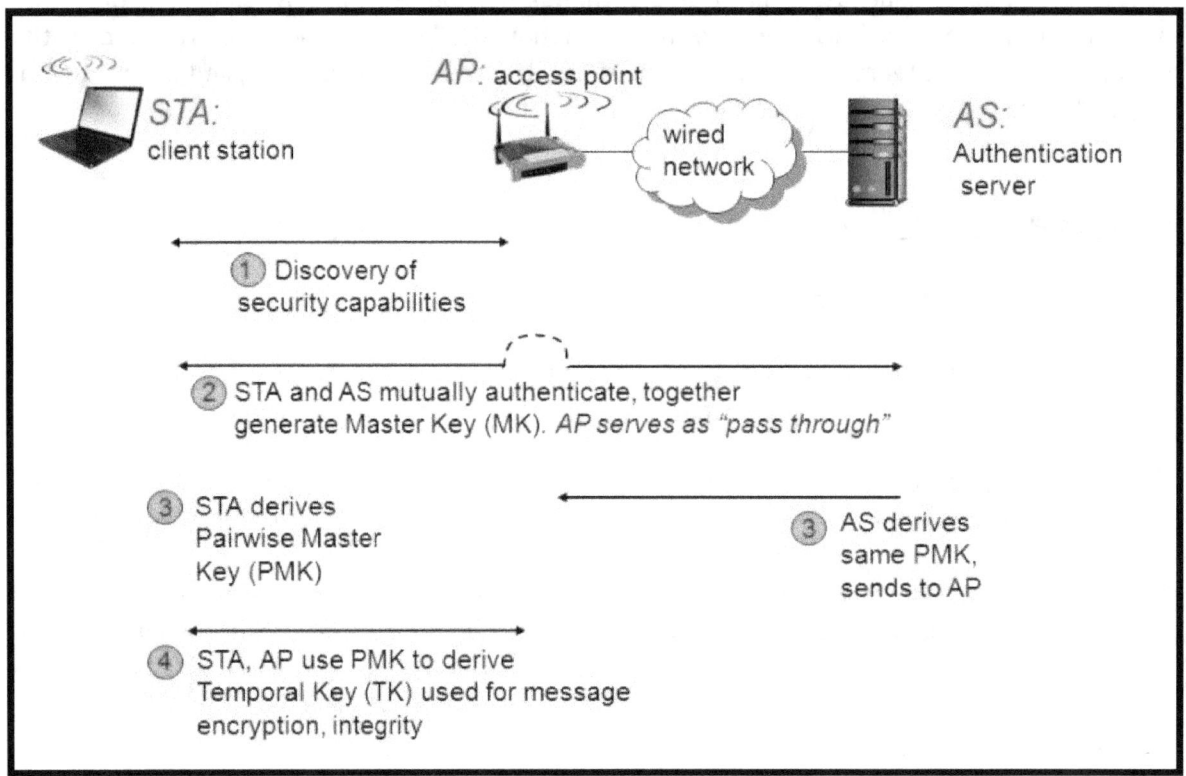

Figure 4.5: Wi-Fi Protected Access (WPA)

Figure 5 gives a general overview of 802.11i operation. First, an exchange between a station and an AP enables the two to agree on a *set of security capabilities* to be used. Then an exchange involving the AS and the station provides for *secure authentication*. The *AS* is responsible for *key distribution to the AP*, which in turn manages and *distributes keys to stations*. Finally, strong encryption is used to protect *data transfer between the station and the AP*.

The 802.11i architecture consists of three main ingredients:

- o *Authentication*: A protocol is used to define an exchange between a user and an AS that provides mutual authentication and generates temporary keys to be used between the client and the AP over the wireless link.

- o *Access Control*: This function enforces the use of the authentication function, routes the messages properly, and facilitates key exchange. It can work with a variety of authentication protocols.

- o *Privacy with Message Integrity*: MAC-level data is encrypted, along with a message integrity code that ensures that the data have not been altered.

4.7 3G versus Wi-Fi:

FUNCTIONS	3G	WI-FI
RADIO INTERFACE	Uses spread spectrum as the modulation technique.	Uses spread spectrum as the modulation technique.
GENESIS	Evolved from voice network where QoS is a critical success factor.	Evolved from data network where QoS is not a critical success factor.
BANDWIDTH	It supports broadband data service of up to 2Mbps.	Wi-Fi supports broadband data service of up to 54Mbps.
STATUS OF STANDARDS	For 3G, there is a relatively small family of internationally sanctioned standards, collectively referred to as IMT-2000.	It is one of the families of continuously evolving 802.11x wireless standards that are under development.
ACCESS TECHNOLOGIES	The wireless link is from the end user device to the cell base station, which may be at a distance of up to a few kilometers.	The wireless link is a few hundred feet from the end-user device to the base station.
BUSINESS MODELS & DEPLOYMENT	Service providers own and manage the infrastructure. End customers typically have a monthly service contract with the 3G service provider to use the network.	User's organization owns the infrastructure. Following the initial investment, the usage of the network does not involve an access fee.
ROAMING	It will offer well-coordinated continuous and ubiquitous coverage.	Seamless ubiquitous roaming over Wi-Fi cannot be guaranteed as network growth is unorganized.

Chapter 5

Introduction to Bluetooth

Bluetooth is an always-on, **short-range radio hookup** that resides on a microchip. It was initially developed by Swedish mobile-phone maker **Ericsson** in **1994**. The Bluetooth standards are published by an industry consortium known as the **Bluetooth SIG (special interest group)**.

The concept behind Bluetooth is to provide a universal short-range wireless capability. Using the **2.4-GHz band**, available globally for **unlicensed low-power uses**, two Bluetooth devices **within 10 m of each other** can share **up to 720 kbps** of capacity.

5.1 Bluetooth Applications:

Bluetooth provides support for three general application areas using **shortrange wireless connectivity**:

- **Data and Voice Access Points**: Bluetooth facilitates real-time voice and data transmissions by providing effortless wireless connection of portable and stationary communications devices.

- **Cable Replacement**: Bluetooth eliminates the need for numerous cable attachments for connection of practically any kind of communication device. Connections are instant and are maintained even when devices are not within line of sight. The range of each radio is approximately 10m but can be extended to 100 m with an optional amplifier.

- **Ad-hoc Networking**: A device equipped with a Bluetooth radio can establish instant connection to another Bluetooth radio as soon as it comes into range.

Following are some of the user scenarios where Bluetooth is used:

1. **Three-in-One Phone**: When you are in the office, your phone functions as an intercom (no telephony charge). At home, it functions as a cordless phone (fixed-line charge). When you are on the move, it functions as a mobile phone (cellular charge).

2. **Internet Bridge**: In this model, a cordless modem acts as a modem to a PC and provides dial-up networking and faxing facilities.

3. **Portable PC Speakerphone:** Connect cordless headsets to your portable PC, and use it as a speaker phone regardless of your location.

4. **Cordless Desktop**: Connect your desktop/laptop computer cordlessly to printers, scanner, keyboard, mouse, and the LAN.

5. **Synchronization:** Automatically synchronize your desktop computer, portable PC, notebook, and mobile phone. As soon as you enter the office, the address list & calendar in your notebook automatically updates the files on your desktop computer or vice versa.

6. **_Briefcase E-mail_:** Access e-mail while your portable PC is still in the briefcase. When your PC receives an e-mail message, you are notified by your mobile phone.

7. **_Interactive Conference_**: In meetings and at conferences, you can share information instantly with other participants. You can also operate a projector remotely without wire connectors.

8. **_The Ultimate Headset_:** Connect a headset to your mobile PC or to any wired connection and free your hands for more important tasks at the office or in your car.

5.2 Piconet & Scatternet in Bluetooth:

The **basic unit of networking** in Bluetooth is a **Piconet**, consisting of **a master** and from **one to seven active slave** devices. The radio designated as the **master makes the determination of the channel** (frequency-hopping sequence) **& phase** (timing offset, i.e., when to transmit) that shall be **used by all devices** on this piconet. A slave may only communicate with the master & may only communicate when granted permission by the master. **Up to eight devices can communicate in a piconet**.

A device in one piconet may also exist as part of another piconet and may function as either a slave or master in each piconet. This form of overlapping is called a **Scatternet**.

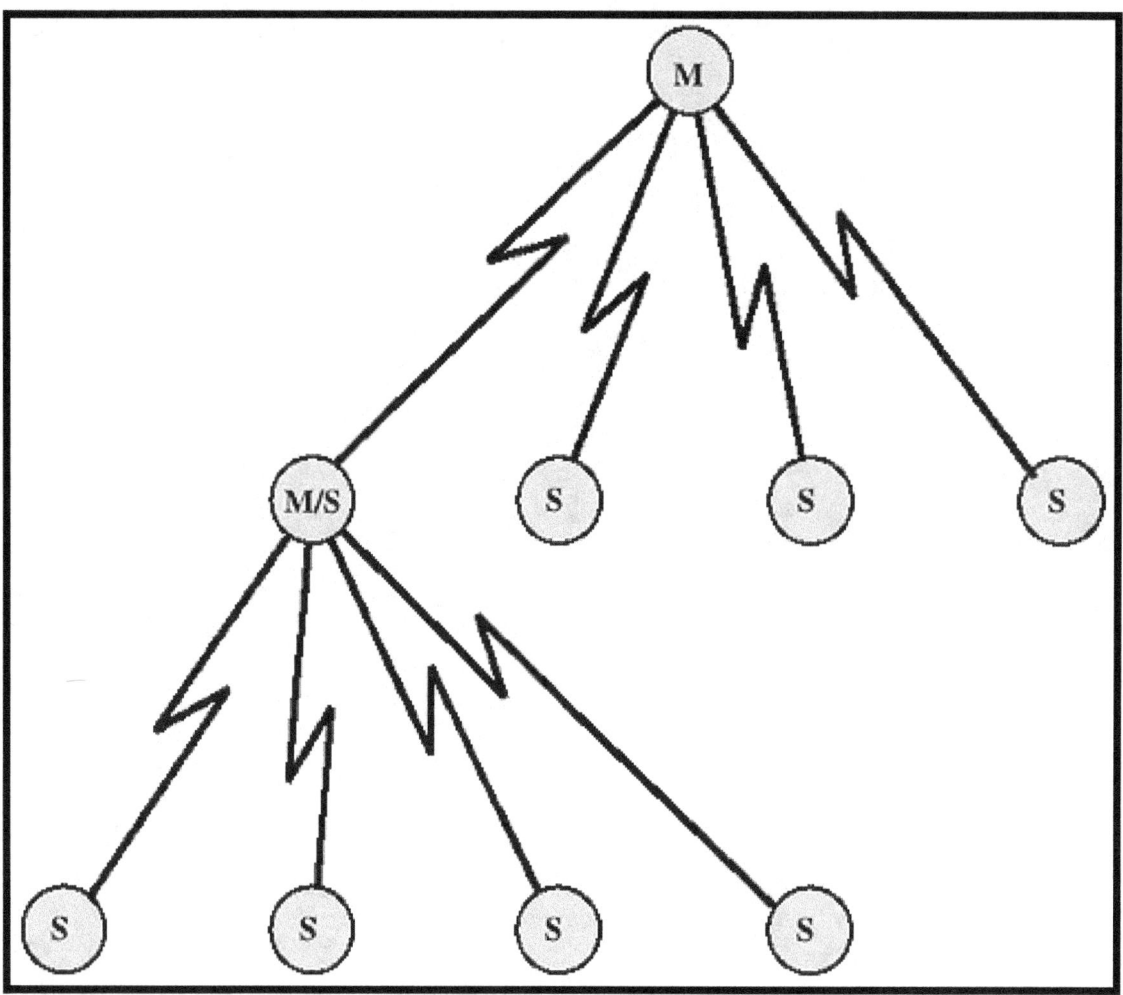

Figure 5.1: Master/Slave Relationships in Scatternet

The advantage of the Piconet/Scatternet scheme is that it allows many devices to share the same physical area and make efficient use of the bandwidth. At any given time, the bandwidth available is **1 MHz**, with a **maximum of eight devices** sharing the bandwidth.

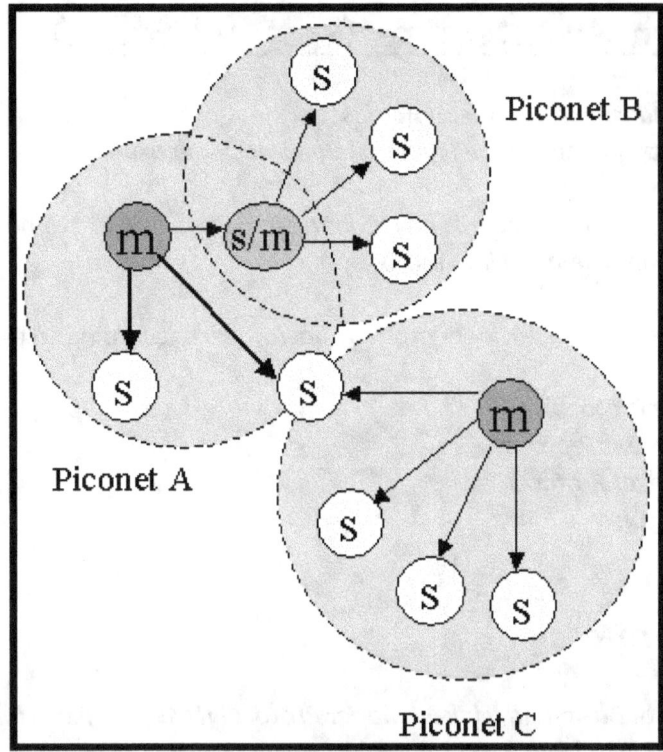

Figure 5.2: Scatternet

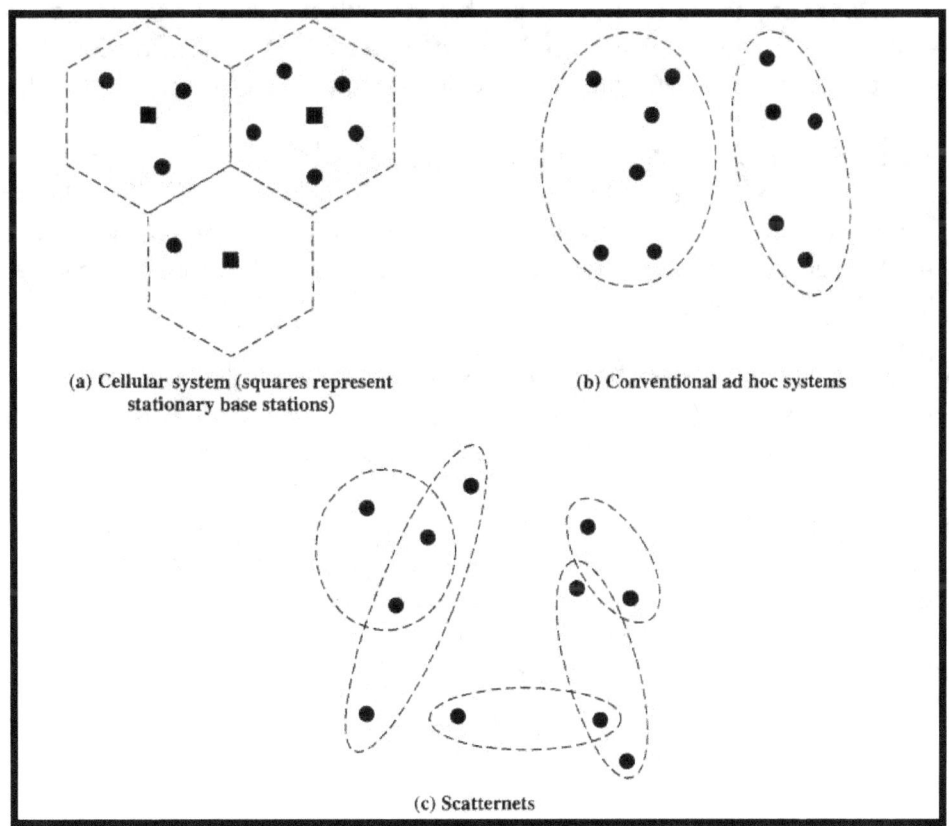

Figure 5.3: Wireless Network Configurations

5.3 Bluetooth Architecture (Protocol Stack):

Bluetooth uses **spread spectrum technologies** at the Physical Layer while using both **direct sequence spread spectrum** and **frequency hopping spread spectrum**.

It uses **connectionless (ACL–Asynchronous Connectionless Link)** & **connection-oriented (SCO–Synchronous Connection-oriented Link)** links.

Bluetooth protocol stack can be divided into **four basic layers** according to their functions:

- **Bluetooth Core Protocols**
- **Cable Replacement Protocol**
- **Telephony Control Protocol**
- **Adopted Protocols**

Bluetooth Core Protocols:

- This comprises of **Baseband**, **Link Manager Protocol (LMP)**, **Logical Link Control & Adaption Protocol (L2CAP)**, and **Service Discovery Protocol (SDP)**.

- **Baseband**: It enables the physical RF link between Bluetooth units forming a piconet. This layer uses inquiry and paging procedures to synchronize the transmission with different Bluetooth devices. Using SCO and ACL link, different packets can be multiplexed.

- **Link Manager Protocol (LMP)**: When two Bluetooth devices come within each other's range, link managers of either device discover each other. LMP then engages itself in peer-to-peer message exchange. These messages perform various security functions starting from authentication to encryption. It also controls the power modes, connection state, and duty cycles of Bluetooth devices in a piconet.

- **Logical Link Control and Adaption Protocol (L2CAP)**: This layer is responsible for segmentation of large packets and the reassembly of fragmented packets. L2CAP is also responsible for multiplexing of Bluetooth packets from different applications.

- **Service Discovery Protocol (SDP)**: It enables a Bluetooth device to join a piconet. Using SDP, a device inquires what services are available in a piconet and how to access them. SDP uses a client-server model where the server has a list of services defined through service records. In Bluetooth device, there is only one SDP server. If a device provides multiple services, one SDP server acts on behalf of all of them.

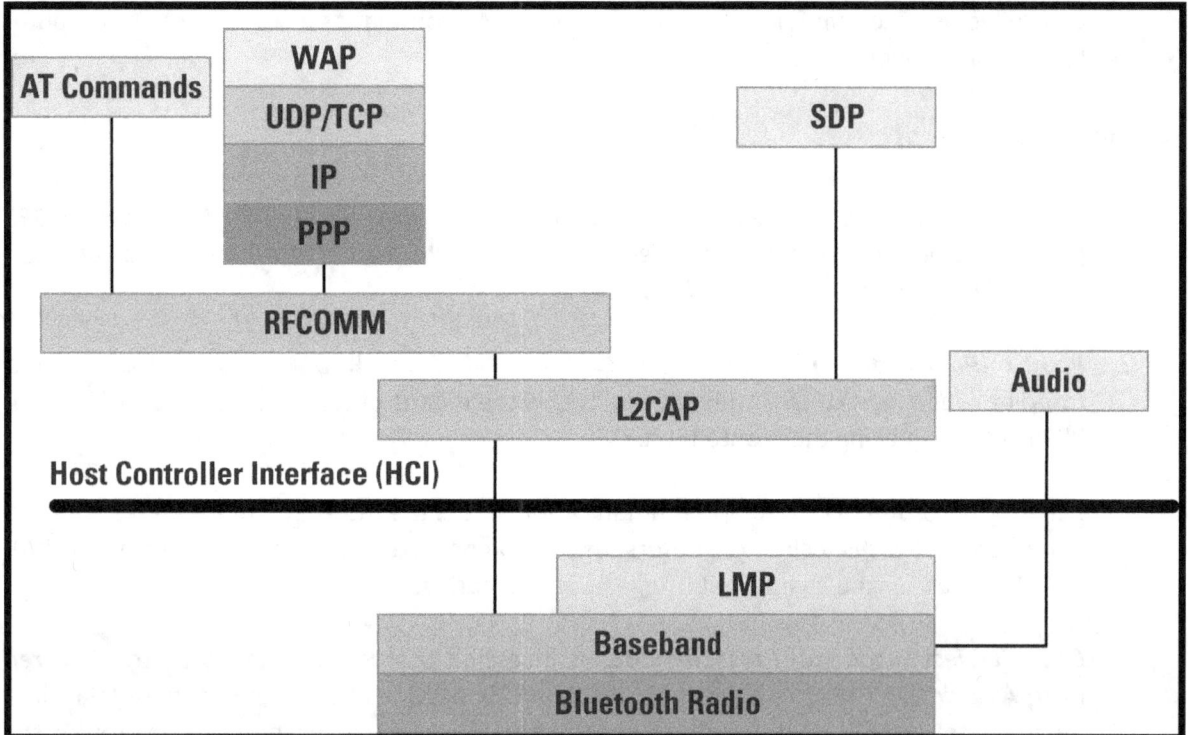

Figure 5.4: Bluetooth Protocol Stack

Cable Replacement Protocol:

- o This protocol has only one member which is *Radio Frequency Communication (RFCOMM)*. *RFCOMM* is a *serial line communication protocol* and is based on ETSI 07.10 specification. The "cable replacement" protocol emulates *RS-232 control and data signals* over Bluetooth Baseband Protocol.

Telephony Control Protocol:

- o It comprises of two protocol stacks, viz., *Telephony Control Specification Binary (TCS BIN)*, and the *AT-commands*.

- o *Telephony Control Specification Binary (TCS BIN)*: It is a bit-oriented protocol. It defines all the call control signaling protocol for set up of speech and data calls between Bluetooth devices. It also defines mobility management procedures for handling groups of Bluetooth devices.

- **AT-Commands**: It defines a set of AT-commands by which a mobile phone can be used and controlled as a modem for fax and data transfers. AT commands are used from a computer to control a modem.

Adopted Protocols:

- This has many protocols stacks like **Point-to-Point Protocol (PPP)**, **TCP/IP Protocol**, **OBEX (Object Exchange Protocol)**, **Wireless Application Protocol (WAP)**, **vCard**, **vCalender**, **Infrared Mobile Communication (IrMC)**, etc.

- **Point-to-Point Protocol (PPP) Bluetooth**: This offers PPP over RFCOMM to accomplish point-to-point connections. Point-to-Point Protocol is the means of taking IP packets to/from the PPP layer and placing them onto the LAN.

- **TCP/IP**: This protocol is used for communication across the Internet. TCP/IP stacks are used in numerous devices including printers, handheld computers, and mobile handsets. TCP/IP or PPP is used for the all Internet bridge usage scenarios.

- **OBEX (Object Exchange) Protocol**: OBEX is a session protocol developed by the **Infrared Data Association** (IrDA) to exchange objects. OBEX provides the functionality of HTTP in a much lighter fashion. It defines a folder listing object, which can be used to browse the contents of folders on remote devices.

- **Content Formats**: **vCard** & **vCalender** specifications define the format of an electronic business card and personal calendar entries developed by the **Versit consortium**. These content formats are used to exchange messages and notes. They are defined in the IrMC specification.

Introduction to Android

6.1 Android Architecture & Application Framework:

Android operating system is a *stack of software components* which is roughly divided into *five sections* and *four main layers* as shown below in the architecture diagram.

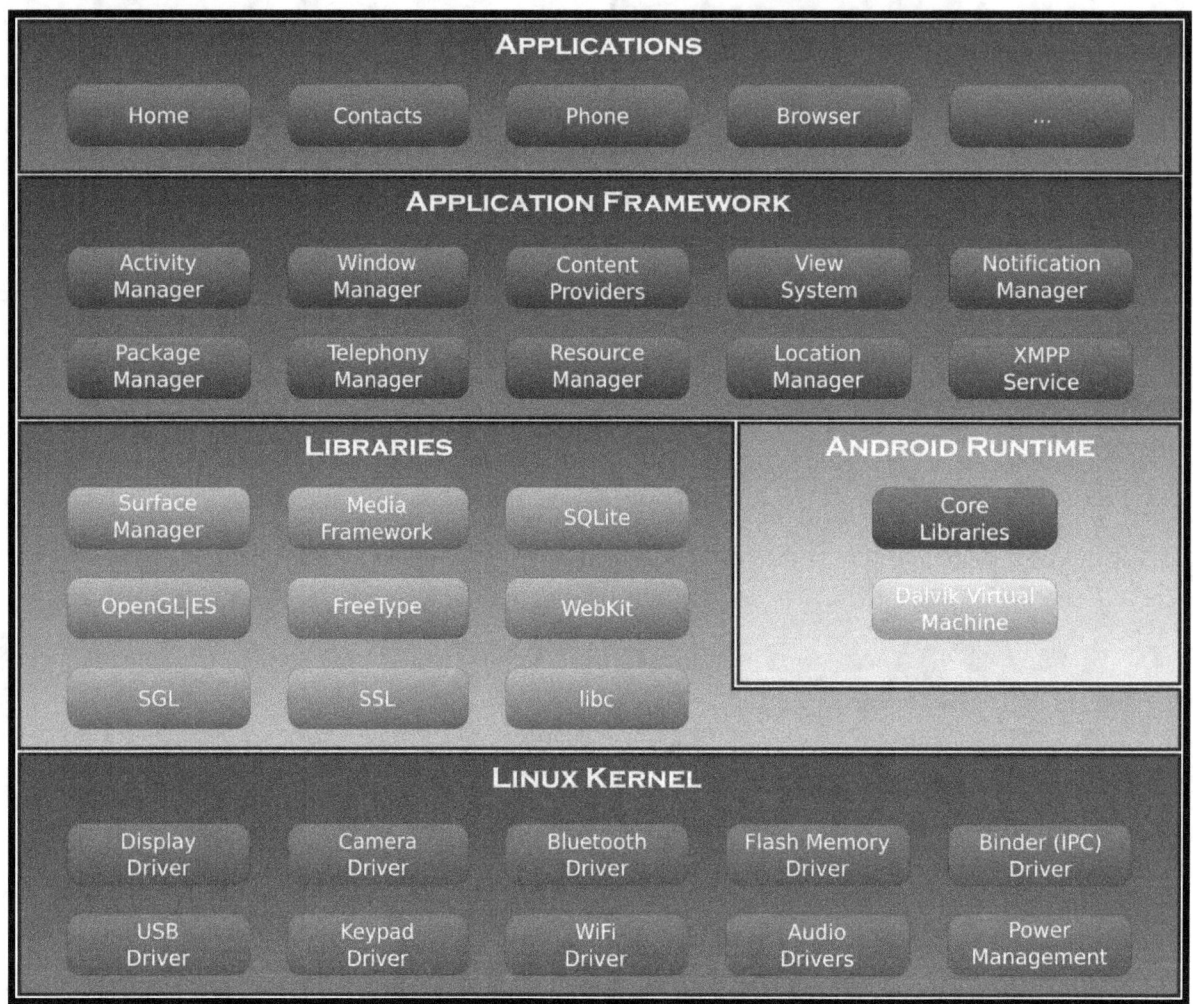

Figure 6.1: Android Architecture

Linux kernel:

This provides a level of **abstraction between the device hardware** and it contains all the essential **hardware drivers** like camera, keypad, display etc. The kernel handles all the things that Linux is really good at such as networking and a **vast array of device drivers**, which take the pain of **interfacing the peripheral hardware**.

Libraries:

On top of Linux kernel there is a **set of libraries** including **open-source Web browser engine - WebKit**, **well known library – libc**, **SQLite database** which is a useful repository for storage and sharing of application data, **libraries to play and record audio and video**, **SSL libraries** responsible for Internet security etc.

Android Runtime:

This section provides a key component called **Dalvik Virtual Machine**, which is a kind of **Java Virtual Machine**, specially designed and optimized **for Android**.

The **Dalvik VM** makes use of **Linux core features** like **memory management** and **multi-threading**, which is intrinsic (built-in) in the Java language. The Dalvik VM **enables every Android application to run in its own process**, **with its own instance of the Dalvik VM**.

The Android runtime also provides a set of core libraries which enable Android application developers to write Android applications using standard Java programming language. A summary of some **key core Android libraries** available to the Android developer is as follows:

1. **android.app** – Provides access to the application model and is the cornerstone of all Android applications.

2. **android.content** – Facilitates content access, publishing and messaging between applications and application components.

3. **android.database** – Used to access data published by content providers and includes SQLite database management classes.

4. **android.opengl** – A Java interface to the OpenGL ES 3D graphics rendering API.

5. **android.os** – Provides applications with access to standard operating system services including messages, system services and inter-process communication.

6. **android.text** – Used to render and manipulate text on a device display.

7. **android.view** – The fundamental building blocks of application user interfaces.

8. **android.widget** – A rich collection of pre-built user interface components such as buttons, labels, list views, layout managers, radio buttons etc.

9. **android.webkit** – A set of classes intended to allow web-browsing capabilities to be built into applications.

Application Framework:

The **Application Framework** layer provides many **higher-level services** to applications in the form of **Java classes**. Application developers are allowed to make use of these services in their applications.

- **Activity Manager** – Controls all aspects of the application lifecycle and activity stack.
- **Content Providers** – Allows applications to publish and share data with other applications.
- **Resource Manager** – Provides access to non-code embedded resources such as strings, color settings and user interface layouts.
- **Notifications Manager** – Allows applications to display alerts and notifications to the user.
- **View System** – an extensible set of views used to create application user interfaces.

Applications:

You will find all the Android application at the **top layer**. You will write your application to be installed on this layer. Examples of such applications are **Contacts**, **Books**, **Browser**, **Games** etc.

Application Components:

Application components are the essential **building blocks of an Android application**. These components are loosely coupled by the application manifest file AndroidManifest.xml that describes each component of the application and how they interact. There are following **four main components that can be used within an Android application**:

1. **Activities:** An activity represents a single screen with a user interface; in-short Activity performs actions on the screen. For example, an email application might have one activity that shows a list of new emails, another activity to compose an email, and another activity for reading emails. If an application has more than one activity, then one of them should be marked as the activity that is presented when the application is launched.

   ```
   public class MainActivity extends Activity {
   }
   ```

2. **Services:** A service is a component that runs in the background to perform long-running operations. For example, a service might play music in the background while the user is in a different application, or it might fetch data over the network without blocking user interaction with an activity.

   ```
   public class MyService extends Service {
   }
   ```

3. **_Broadcast Receivers:_** Broadcast Receivers simply respond to broadcast messages from other applications or from the system. For example, applications can also initiate broadcasts to let other applications know that some data has been downloaded to the device and is available for them to use, so this is broadcast receiver who will intercept this communication and will initiate appropriate action. A broadcast receiver is implemented as a subclass of BroadcastReceiver class.

   ```
   public class MyReceiver extends BroadcastReceiver {
         public void onReceive(context,intent) {
         }
   }
   ```

4. **_Content Providers_**: A content provider component supplies data from one application to others on request. Such requests are handled by the methods of the ContentResolver class. The data may be stored in the file system, the database or somewhere else entirely. A content provider is implemented as a subclass of ContentProvider class and must implement a standard set of APIs that enable other applications to perform transactions.

   ```
   public class MyContentProvider extends ContentProvider {
         public void onCreate() {
         }
   }
   ```

6.2 The Manifest File:

Every application must have an AndroidManifest.xml file in its root directory. The manifest file provides essential information about your app to the Android system, which the system must have before it can run any of the app's code.

Manifest file structure

```
<?xml version="1.0" encoding="utf-8"?>
<manifest>

  <uses-permission />
  <permission />
  <permission-tree />
  <permission-group />
  <instrumentation />
  <uses-sdk />
  <uses-configuration />
  <uses-feature />
  <supports-screens />
  <compatible-screens />
  <supports-gl-texture />

  <application>

    <activity>
      <intent-filter>
        <action />
        <category />
        <data />
      </intent-filter>
      <meta-data />
    </activity>

    <activity-alias>
      <intent-filter> . . . </intent-filter>
      <meta-data />
    </activity-alias>

    <service>
      <intent-filter> . . . </intent-filter>
      <meta-data/>
    </service>

    <receiver>
      <intent-filter> . . . </intent-filter>
      <meta-data />
    </receiver>

    <provider>
      <grant-uri-permission />
      <meta-data />
      <path-permission />
    </provider>

    <uses-library />

  </application>

</manifest>
```

1. **<manifest> :**

    <manifest   xmlns:android=http://schemas.android.com/apk/res/android
                 package="string"
                 android:versionCode="integer"
                 android:versionName="string">
        . . .
    </manifest>

It's the root element of the AndroidManifest.xml file. It must contain an <application> element and specify xmlns:android and package attributes.

2. **<application> :**

   <application   android:allowBackup=["true" | "false"]
                  android:icon="drawable resource"
                  android:label="string resource"
                  android:logo="drawable resource"
                  android:name="string">

   . . .
   </application>

   It's the declaration of the application. This element contains sub elements that declare each of the application's components and has attributes that can affect all the components.

3. **<activity> :**

   <activity   android:icon="drawable resource"
                  android:label="string resource"
                  android:name="string"
                  android:theme="resource or theme">

   . . .
   </activity>

   Declares an activity that implements part of the application's visual user interface. All activities must be represented by <activity> elements in the manifest file. Any that are not declared there will not be seen by the system and will never be run.

4. **<action> :**

   <action android:name="string" />

   Adds an action to an intent filter. An <intent-filter> element must contain one or more <action> elements. If it doesn't contain any, no Intent objects will get through the filter.

5. **<service> :**

    ```
    <service     android:icon="drawable resource"
                 android:label="string resource"
                 android:name="string"
                 android:process="string" >
            . . .
    </service>
    ```

    They're used to implement long-running background operations or a rich communications API that can be called by other applications. All services must be represented by <service> elements in the manifest file. Any that are not declared there will not be seen by the system and will never be run.

6.3 Android Layouts:

The **basic building block** for user interface is a **View object** which is created from the **View class**. It occupies a rectangular area on the screen and is responsible for **drawing and event handling**.

View is the **base class for widgets**, which are used to **create interactive UI components** like buttons, text fields, etc.

The **ViewGroup** is a subclass of View and **provides invisible container** that **hold other Views** or other ViewGroups and define their layout properties.

6.3.1 LinearLayout:

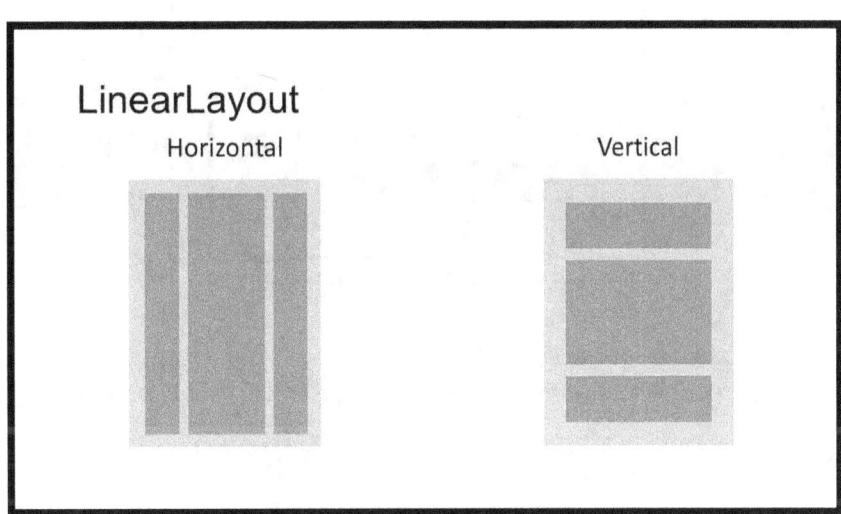

Figure 6.2: Linear Layout – Android

In a linear layout, like the name suggests, all the elements are displayed in a linear fashion, either Horizontally or Vertically and this behavior is set in android:orientation which is an attribute of the node LinearLayout.

```
<?xml version="1.0" encoding="utf-8"?>
<LinearLayout    xmlns:android="http://schemas.android.com/apk/res/android"
                 android:orientation="vertical OR horizontal"
                 android:layout_width="match_parent"
                 android:layout_height="match_parent">

    ...
</LinearLayout>
```

6.3.2 RelativeLayout:

In a relative layout, every element arranges itself relative to other elements or a parent element. As an example, let's consider the layout defined below.

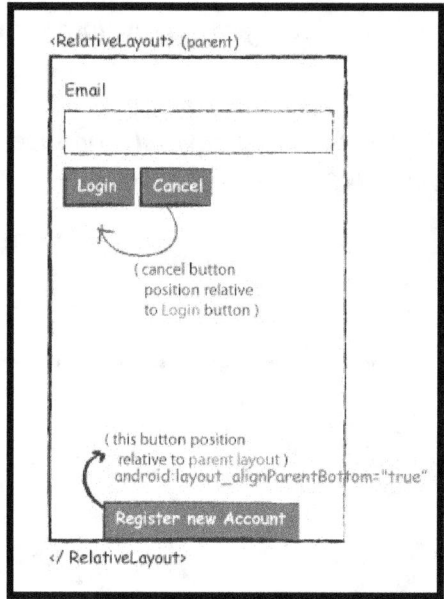

Figure 6.3: Relative Layout - Android

The "Cancel" button is placed relatively, to the right of the "Login" button parallelly. Here is the code snippet that achieves the mentioned alignment (Right of Login button parallelly).

```
<Button android:id="@+id/btnLogin" ...></Button>

<Button android:layout_toRightOf="@id/btnLogin"
        android:layout_alignTop="@id/btnLogin" ...></Button>
```

android:layout_alignParentTop
　　　　If "true", makes the top edge of this view match the top edge of the parent.

android:layout_centerVertical
　　　　If "true", centers this child vertically within its parent.

android:layout_below
　　　　Positions the top edge of this view below the view specified with a resource ID.

android:layout_toRightOf
　　　　Positions the left edge of this view to the right of the view specified with a resource ID.

6.3.3 *ScrollView:*

ScrollView is a special kind of layout, designed to hold view larger than its actual size. When the Views size goes beyond the ScrollView size, it automatically adds scroll bars and can be scrolled vertically.

ScrollView can hold only one direct child. This means that, if you have complex layout with more view controls, you must enclose them inside another standard layout like LinearLayout, TableLayout or RelativeLayout.

You should never use a ScrollView with a ListView or GridView, because they both takes care of their own vertical scrolling.

ScrollView only supports vertical scrolling. Use HorizontalScrollView for horizontal scrolling.

```xml
<ScrollView  xmlns:android="http://schemas.android.com/apk/res/android"
             android:layout_width="match_parent"
             android:layout_height="wrap_content"
             android:orientation="vertical">

    <LinearLayout android:layout_width="match_parent"
                  android:layout_height="wrap_content"
                  android:orientation="vertical">

        <ImageView   android:id="@+id/imageView"
                     android:layout_width="wrap_content"
                     android:layout_height="200dp"
                     android:scaleType="centerCrop"
                     android:src="@drawable/image" />

        <TextView    android:id="@+id/textView"
                     android:layout_width="wrap_content"
                     android:layout_height="wrap_content"
                     android:text="@string/description"/>

    </LinearLayout>

</ScrollView>
```

6.3.4 TableLayout:

TableLayout in Android arranges a group of views into rows and columns.

You will use the <TableRow> element to build a row in the table. Each row has zero or more cells; each cell can hold one View object.

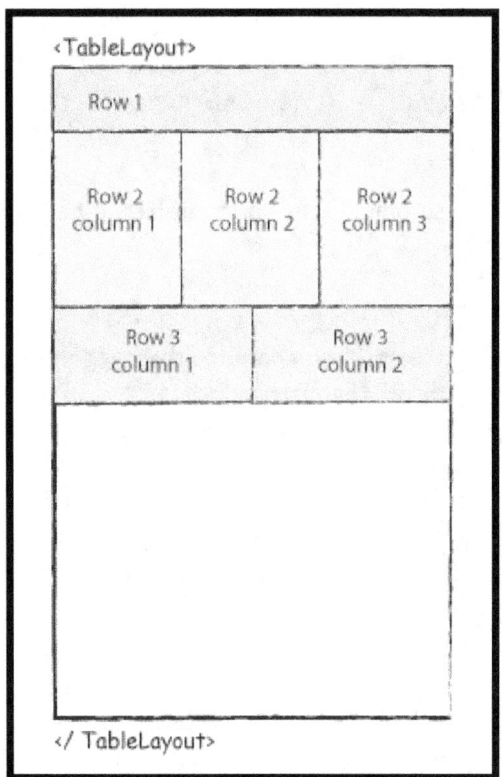

Figure 6.4: Table Layout - Android

```
<TableLayout xmlns:android="http://schemas.android.com/apk/res/android"
        android:layout_width="match_parent"
        android:layout_height="match_parent"
        android:shrinkColumns="*"
        android:stretchColumns="*"
        android:background="#ffffff">

<!-- Row 1 with single column -->
    <TableRow android:layout_height="wrap_content"
            android:layout_width="fill_parent"
            android:gravity="center_horizontal">

        <TextView  android:layout_width="match_parent"
                android:layout_height="wrap_content"
                android:textSize="18dp"
```

```xml
                    android:text="Row 1"
                    android:layout_span="3"
                    android:padding="18dip"
                    android:background="#b0b0b0"
                    android:textColor="#000"/>
    </TableRow>

<!-- Row 2 with 3 columns -->
    <TableRow    android:id="@+id/tableRow1"
                android:layout_height="wrap_content"
                android:layout_width="match_parent">
        <TextView
            android:id="@+id/TextView04" android:text="Row 2 column 1"
            android:layout_weight="1" android:background="#dcdcdc"
            android:textColor="#000000"
            android:padding="20dip" android:gravity="center"/>
        <TextView
            android:id="@+id/TextView04" android:text="Row 2 column 2"
            android:layout_weight="1" android:background="#d3d3d3"
            android:textColor="#000000"
            android:padding="20dip" android:gravity="center"/>
        <TextView
            android:id="@+id/TextView04" android:text="Row 2 column 3"
            android:layout_weight="1" android:background="#cac9c9"
            android:textColor="#000000"
            android:padding="20dip" android:gravity="center"/>
    </TableRow>

<!-- Row 3 with 2 columns -->
    <TableRow
        android:layout_height="wrap_content"
        android:layout_width="fill_parent"
        android:gravity="center_horizontal">
        <TextView
            android:id="@+id/TextView04" android:text="Row 3 column 1"
            android:layout_weight="1" android:background="#b0b0b0"
            android:textColor="#000000"
            android:padding="20dip" android:gravity="center"/>

        <TextView
            android:id="@+id/TextView04" android:text="Row 3 column 2"
            android:layout_weight="1" android:background="#a09f9f"
            android:textColor="#000000"
            android:padding="20dip" android:gravity="center"/>
    </TableRow>

</TableLayout>
```

6.3.5 FrameLayout:

Frame Layout is designed to block out an area on the screen to display a single item. Generally, FrameLayout should be used to hold a single child view, because it can be difficult to organize child views in a way that's scalable to different screen sizes without the children overlapping each other.

```xml
<FrameLayout    xmlns:android="http://schemas.android.com/apk/res/android"
                android:layout_width="fill_parent"
                android:layout_height="fill_parent">

  <ImageView
    android:src="@drawable/ic_launcher"
    android:scaleType="fitCenter"
    android:layout_height="250px"
    android:layout_width="250px"/>

  <TextView
    android:text="Frame Demo"
    android:textSize="30px"
    android:textStyle="bold"
    android:layout_height="fill_parent"
    android:layout_width="fill_parent"
    android:gravity="center"/>

</FrameLayout>
```

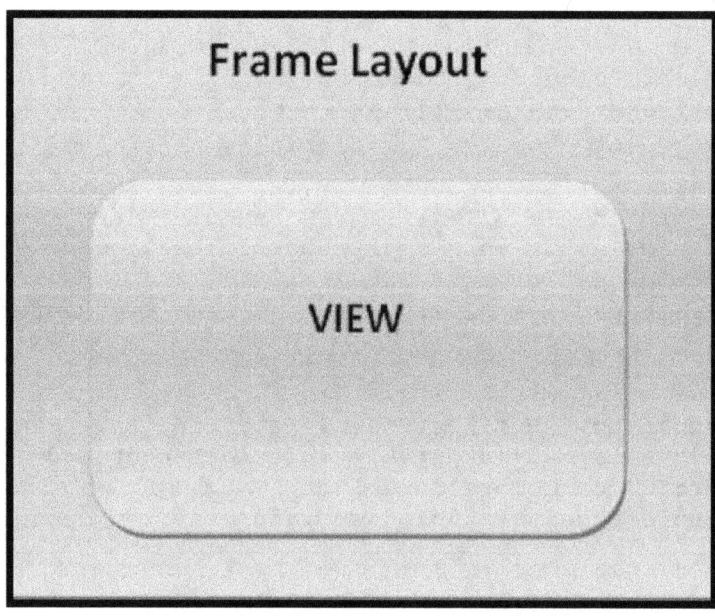

Figure 6.5: Frame Layout – Android

6.4 Using the TextView:

A TextView displays text to the user and optionally allows them to edit it. A TextView is a complete text editor, however the basic class is configured to not allow editing.

Main Attributes of Text view:

android:gravity
Specifies how to align the text by the view's x- and/or y-axis when the text is smaller than the view.

android:hint
Hint text to display when the text is empty.

android:id
This is the ID which uniquely identifies the control.

android:text
Text to display.

android:textColor
Text color. May is a color value, in the form of "#rgb", "#argb", "#rrggbb", or "#aarrggbb".

android:textSize
Size of the text. Recommended dimension type for text is "sp" for scaled-pixels (example: 15sp)

Example of TextView:

```
<RelativeLayout xmlns:android="http://schemas.android.com/apk/res/android"
  xmlns:tools="http://schemas.android.com/tools"
  android:layout_width="match_parent"
  android:layout_height="match_parent"
  android:paddingBottom="@dimen/activity_vertical_margin"
  android:paddingLeft="@dimen/activity_horizontal_margin"
  android:paddingRight="@dimen/activity_horizontal_margin"
  android:paddingTop="@dimen/activity_vertical_margin"
  tools:context=".MainActivity" >

  <TextView
    android:id="@+id/text_id"
    android:layout_width="300dp"
    android:layout_height="200dp"
    android:capitalize="characters"
    android:text="hello_world"
```

```
        android:textColor="@android:color/holo_blue_dark"
        android:textColorHighlight="@android:color/primary_text_dark"
        android:layout_centerVertical="true"
        android:layout_alignParentEnd="true"
        android:textSize="50dp"/>
</RelativeLayout>
```

Figure 6.6: TextView - Android

6.5 EditText View:

EditText is an overlay over TextView that configures itself to be editable. It is the predefined subclass of TextView that includes rich editing capabilities.

Edittext Attributes:

android:text
If set, specifies that this TextView has a textual input method and automatically corrects some common spelling errors.

android:background
This is a drawable to use as the background.

android:id
This supplies an identifier name for this view.

android:visibility
This controls the initial visibility of the view.

Example of EditText View:

```xml
<RelativeLayout xmlns:android="http://schemas.android.com/apk/res/android"
  xmlns:tools="http://schemas.android.com/tools"
  android:layout_width="match_parent"
  android:layout_height="match_parent"
  tools:context=".MainActivity" >

  <EditText
    android:id="@+id/edittext"
    android:layout_width="fill_parent"
    android:layout_height="wrap_content"
    android:layout_alignLeft="@+id/button"
    android:layout_below="@+id/textView1"
    android:layout_marginTop="61dp"
    android:ems="10"
    android:text="@string/enter_text" android:inputType="text" />

</RelativeLayout>
```

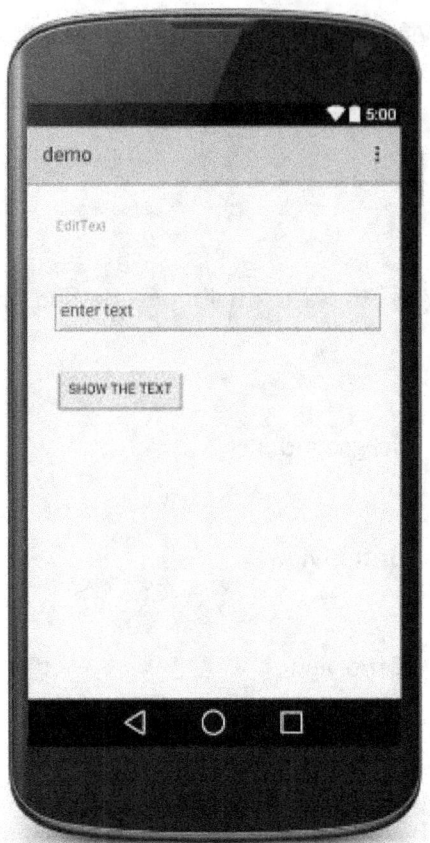

Figure 6.7: EditText View – Android

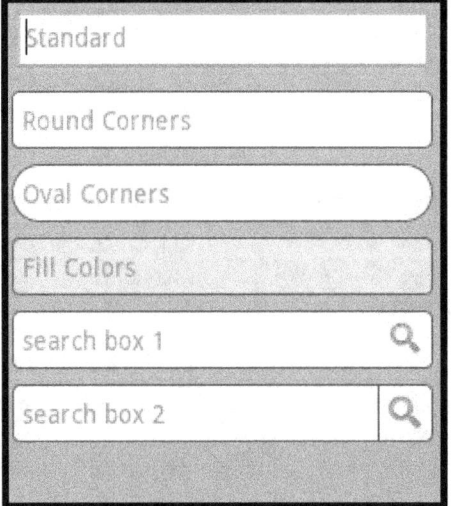

Figure 6.8: EditText View - Styles - Android

6.6 Button View:

A Button is a Push-button which can be pressed, or clicked, by the user to perform an action.

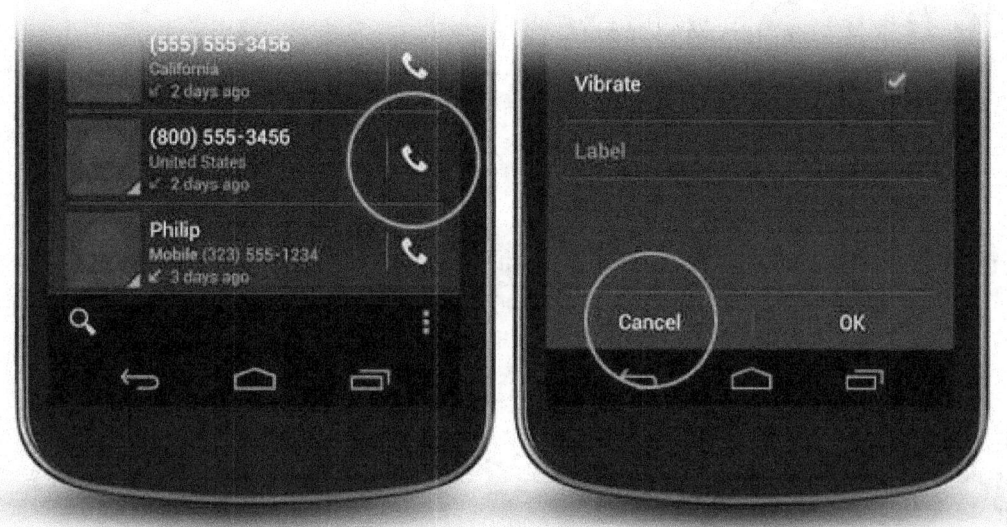

Figure 6.9: Button in Android

Button Attributes:

android:drawableBottom
This is the drawable to be drawn below the text.

android:text
This is the Text to display.

android:background
This is a drawable to use as the background.

android:id
This supplies an identifier name for this view.

android:id
This supplies an identifier name for this view.

Example of Button:

```xml
<?xml version="1.0" encoding="utf-8"?>
<RelativeLayout
  xmlns:android="http://schemas.android.com/apk/res/android"
  xmlns:tools="http://schemas.android.com/tools"
  android:layout_width="match_parent"
  android:layout_height="match_parent"
  android:paddingLeft="@dimen/activity_horizontal_margin"
  android:paddingRight="@dimen/activity_horizontal_margin"
  android:paddingTop="@dimen/activity_vertical_margin"
  android:paddingBottom="@dimen/activity_vertical_margin"
  tools:context=".MainActivity">

  <Button
    android:layout_width="wrap_content"
    android:layout_height="wrap_content"
    android:text="Button"
    android:id="@+id/button"
    android:layout_alignTop="@+id/editText"
    android:layout_alignLeft="@+id/textView1"
    android:layout_alignStart="@+id/textView1"
    android:layout_alignRight="@+id/editText"
    android:layout_alignEnd="@+id/editText" />

</RelativeLayout>
```

6.7 RadioButton:

A RadioButton has two states: either checked or unchecked. This allows the user to select one option from a set. For Using RadioButton we use RadioGroup Control. If we check one RadioButton that belongs to a radio group, it automatically unchecks any previously checked radio button within the same group.

Figure 6.10: RadioButton – Android

Radio Group Attributes:

android:checkedButton
This is the id of child radio button that should be checked by default within this radio group.

android:background
This is a drawable to use as the background.

android:id
This supplies an identifier name for this view.

android:visibility
This controls the initial visibility of the view.

android:onClick
This is the name of the method in this View's context to invoke when the view is clicked.

Example of RadioButton:

```
<?xml version="1.0" encoding="utf-8"?>

<RelativeLayout xmlns:android="http://schemas.android.com/apk/res/android"
    xmlns:tools="http://schemas.android.com/tools"
    android:layout_width="match_parent"
    android:layout_height="match_parent"
    tools:context=".MainActivity">
```

```xml
<RadioGroup
    android:id="@+id/radioGroup"
    android:layout_width="fill_parent"
    android:layout_height="fill_parent"
    android:layout_below="@+id/imageButton"
    android:layout_alignLeft="@+id/textView2"
    android:layout_alignStart="@+id/textView2">

    <RadioButton
        android:layout_width="142dp"
        android:layout_height="wrap_content"
        android:text="JAVA"
        android:id="@+id/radioButton"
        android:textSize="25dp"
        android:textColor="@android:color/holo_red_light"
        android:checked="false"
        android:layout_gravity="center_horizontal" />

    <RadioButton
        android:layout_width="wrap_content"
        android:layout_height="wrap_content"
        android:text="ANDROID"
        android:id="@+id/radioButton2"
        android:layout_gravity="center_horizontal"
        android:checked="false"
        android:textColor="@android:color/holo_red_dark"
        android:textSize="25dp" />

    <RadioButton
        android:layout_width="136dp"
        android:layout_height="wrap_content"
        android:text="HTML"
        android:id="@+id/radioButton3"
        android:layout_gravity="center_horizontal"
        android:checked="false"
        android:textSize="25dp"
        android:textColor="@android:color/holo_red_dark" />
</RadioGroup>
</RelativeLayout>
```

6.8 CheckBox:

A CheckBox is an on/off switch that can be toggled by the user. You should use check-boxes when presenting users with a group of selectable options that are not mutually exclusive.

CheckBox Attributes:

android:autoText
If set, specifies that this TextView has a textual input method and automatically corrects some common spelling errors.

android:drawableRight
This is the drawable to be drawn to the right of the text.

android:text
This is the Text to display. Inherited from android.view.View Class

android:background
This is a drawable to use as the background.

android:contentDescription
This defines text that briefly describes content of the view.

android:id
This supplies an identifier name for this view.

android:onClick
This is the name of the method in this View's context to invoke when the view is clicked.

android:visibility
This controls the initial visibility of the view.

Example of CheckBox:

```xml
<RelativeLayout
    xmlns:android="http://schemas.android.com/apk/res/android"
    xmlns:tools="http://schemas.android.com/tools"
    android:layout_width="match_parent"
    android:layout_height="match_parent"
    tools:context=".MainActivity">
```

```xml
<CheckBox
    android:id="@+id/checkBox1"
    android:layout_width="wrap_content"
    android:layout_height="wrap_content"
    android:text="Do you like Java"
    android:layout_above="@+id/button"
    android:layout_centerHorizontal="true" />

<CheckBox
    android:id="@+id/checkBox2"
    android:layout_width="wrap_content"
    android:layout_height="wrap_content"
    android:text="Do you like android "
    android:checked="false"
    android:layout_above="@+id/checkBox1"
    android:layout_alignLeft="@+id/checkBox1"
    android:layout_alignStart="@+id/checkBox1" />

</RelativeLayout>
```

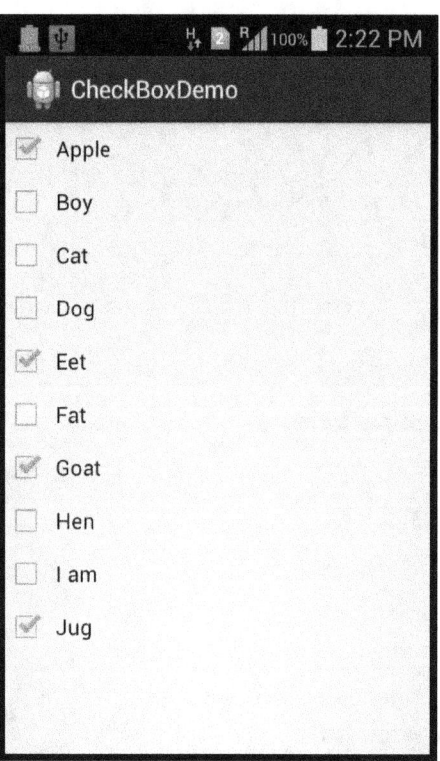

Figure 6.11: Checkbox – Android

6.9 ImageButton:

An ImageButton shows a button with an image (instead of text) that can be pressed or clicked by the user.

ImageButton Attributes:

android:adjustViewBounds
Set this to true if you want the ImageView to adjust its bounds to preserve the aspect ratio of its drawable.

android:baseline
This is the offset of the baseline within this view.

android:baselineAlignBottom
If true, the image view will be baseline aligned with based on its bottom edge.

android:cropToPadding
If true, the image will be cropped to fit within its padding.

android:src
This sets a drawable as the content of this ImageView.

android:id
This supplies an identifier name for this view.

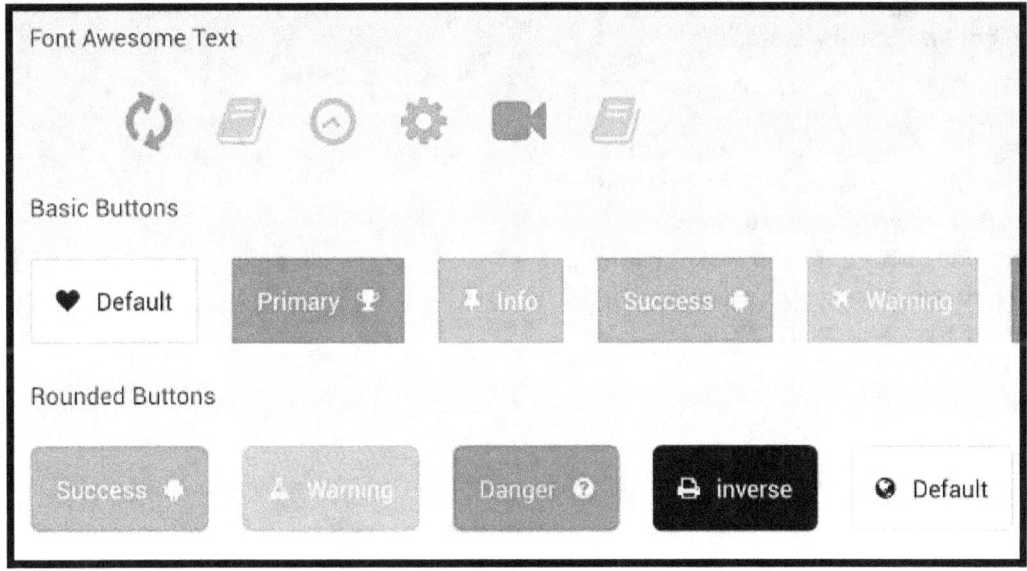

Figure 6.12: ImageButton – Styles - Android

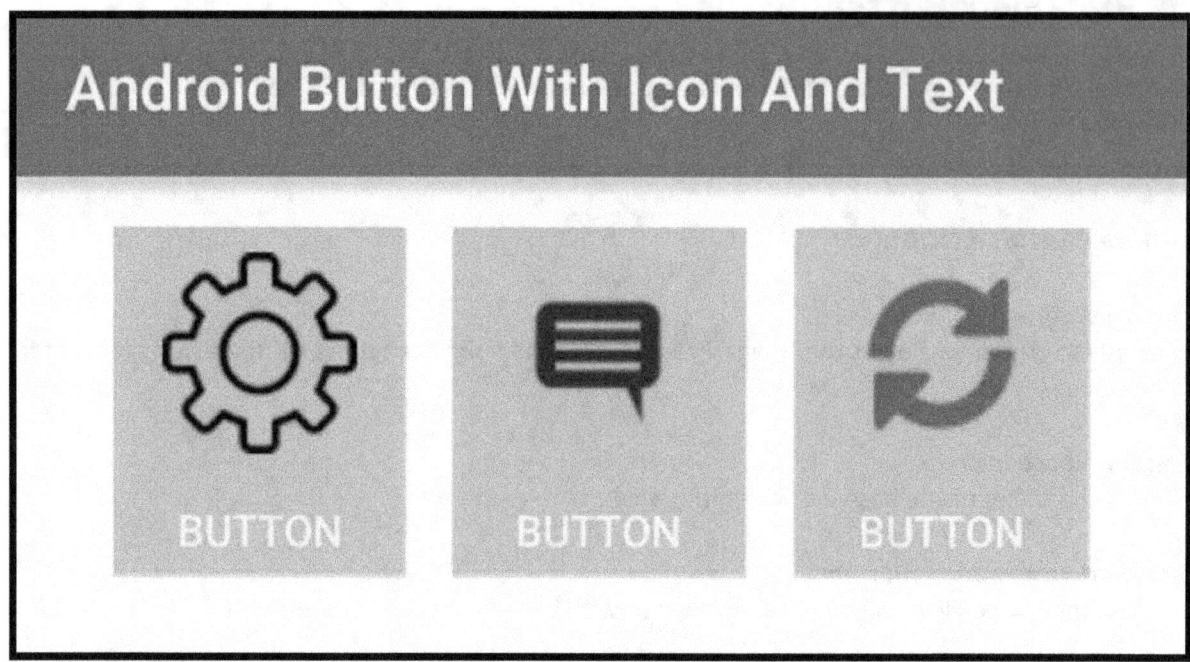

Figure 6.13: ImageButton – Android

Example of ImageButton:

```
<?xml version="1.0" encoding="utf-8"?>
<RelativeLayout
    xmlns:android="http://schemas.android.com/apk/res/android"
    xmlns:tools="http://schemas.android.com/tools" android:layout_width="match_parent"
    android:layout_height="match_parent"
    tools:context=".MainActivity">

    <ImageButton
        android:layout_width="wrap_content"
        android:layout_height="wrap_content"
        android:id="@+id/imageButton"
        android:layout_centerVertical="true"
        android:layout_centerHorizontal="true"
        android:src="@drawable/abc"/>

</RelativeLayout>
```

6.10 RatingBar:

Android RatingBar can be used to get the rating from the user. The Rating returns a floating-point number. It may be 2.0, 3.5, 4.0 etc. Android RatingBar displays the rating in stars. Android RatingBar is the subclass of AbsSeekBar class. The getRating() method of android RatingBar class returns the rating number.

Example of RatingBar:

```
<RelativeLayout xmlns:androclass="http://schemas.android.com/apk/res/android"
  xmlns:tools="http://schemas.android.com/tools"
  android:layout_width="match_parent"
  android:layout_height="match_parent"
  tools:context=".MainActivity" >

  <RatingBar
    android:id="@+id/ratingBar1"
    android:layout_width="wrap_content"
    android:layout_height="wrap_content"
    android:layout_alignParentTop="true"
    android:layout_centerHorizontal="true"
    android:layout_marginTop="44dp" />

  <Button
    android:id="@+id/button1"
    android:layout_width="wrap_content"
    android:layout_height="wrap_content"
    android:layout_alignLeft="@+id/ratingBar1"
    android:layout_below="@+id/ratingBar1"
    android:layout_marginLeft="92dp"
    android:layout_marginTop="66dp"
    android:text="submit" />

</RelativeLayout>
```

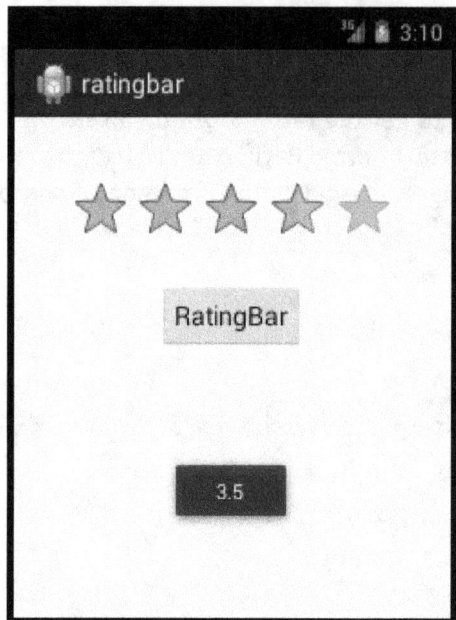

Figure 6.14: RatingBar – Android

6.11 The ProgressBar:

Progress bars are used to show progress of a task. For example, when you are uploading or downloading something from the internet, it is better to show the progress of download/upload to the user.

In Android, there is a class called ProgressDialog that allows you to create progress bar. In order to do this, you need to instantiate an object of this class. Its syntax is:

ProgressDialog progress = new ProgressDialog(this);

Now you can set some properties of this dialog. Such as, its style, its text etc.

progress.setMessage("Downloading Music :) ");
progress.setProgressStyle(ProgressDialog.STYLE_HORIZONTAL);
progress.setIndeterminate(true);

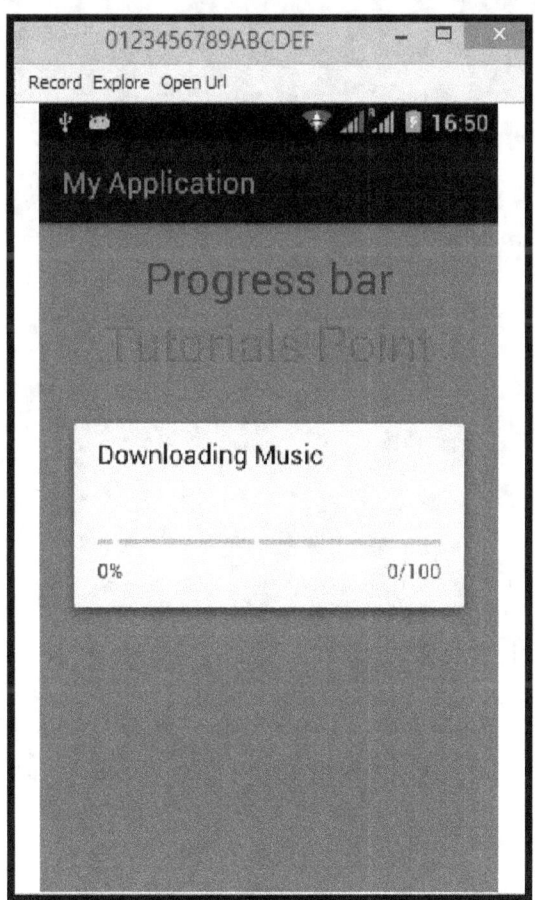

Figure 6.15: ProgressBar – Android

Example of ProgressBar (src/MainActivity.java):

```java
package com.example.mohit.myapplication;

import android.app.ProgressDialog;
import android.support.v7.app.ActionBarActivity;
import android.os.Bundle;
import android.view.View;
import android.widget.Button;

public class MainActivity extends ActionBarActivity {
  Button b1;
  private ProgressDialog progress;
    protected void onCreate(Bundle savedInstanceState) {
    super.onCreate(savedInstanceState);
    setContentView(R.layout.activity_main);
    b1 = (Button) findViewById(R.id.button2);
  }

  public void download(View view){
    progress=new ProgressDialog(this);
    progress.setMessage("Downloading Music");
    progress.setProgressStyle(ProgressDialog.STYLE_HORIZONTAL);
    progress.setIndeterminate(true);
    progress.setProgress(0);
    progress.show();
      final int totalProgressTime = 100;
    final Thread t = new Thread() {
     @Override
     public void run() {
      int jumpTime = 0;

      while(jumpTime < totalProgressTime) {
        try {
          sleep(200);
          jumpTime += 5;
          progress.setProgress(jumpTime);
        } catch (InterruptedException e) {
          // TODO Auto-generated catch block
          e.printStackTrace();
        }
       }
      }
     };
    t.start();
  }
}
```

(res/layout/activity_main.xml):

```xml
<RelativeLayout xmlns:android="http://schemas.android.com/apk/res/android"
    xmlns:tools="http://schemas.android.com/tools"
    android:layout_width="match_parent"
    android:layout_height="match_parent"
    tools:context=".MainActivity">

    <Button
        android:layout_width="wrap_content"
        android:layout_height="wrap_content"
        android:text="Download"
        android:onClick="download"
        android:id="@+id/button2"
        android:layout_marginLeft="125dp"
        android:layout_marginStart="125dp"
        android:layout_centerVertical="true" />

</RelativeLayout>
```

6.12 The Context Menu:

Android context menu appears when user press long click on the element. It is also known as floating menu.

Example of Context Menu:

activity_main.xml:

```xml
<RelativeLayout xmlns:android="http://schemas.android.com/apk/res/android"
  xmlns:tools="http://schemas.android.com/tools"
  android:layout_width="match_parent"
  android:layout_height="match_parent"
  android:paddingBottom="@dimen/activity_vertical_margin"
  android:paddingLeft="@dimen/activity_horizontal_margin"
  android:paddingRight="@dimen/activity_horizontal_margin"
  android:paddingTop="@dimen/activity_vertical_margin"
  tools:context=".MainActivity" >

  <ListView
    android:id="@+id/listView1"
    android:layout_width="match_parent"
    android:layout_height="wrap_content"
    android:layout_alignParentLeft="true"
    android:layout_alignParentTop="true"
    android:layout_marginLeft="66dp"
    android:layout_marginTop="53dp" >
  </ListView>

</RelativeLayout>
```

MainActivity.java:

```java
package com.javatpoint.contextmenu;
import android.os.Bundle;
import android.app.Activity;
import android.view.ContextMenu;
import android.view.ContextMenu.ContextMenuInfo;
import android.view.Menu;
import android.view.MenuItem;
import android.view.View;
import android.widget.AdapterView;
import android.widget.ArrayAdapter;
import android.widget.ListView;
import android.widget.Toast;
```

```java
public class MainActivity extends Activity {
  ListView listView1;
  String contacts[]={"Ajay","Sachin","Sumit","Tarun","Yogesh"};
  @Override
  protected void onCreate(Bundle savedInstanceState) {
    super.onCreate(savedInstanceState);
    setContentView(R.layout.activity_main);
    listView1=(ListView)findViewById(R.id.listView1);
    ArrayAdapter<String> adapter=new ArrayAdapter<String>(this,android.R.layout.simple_list_item_1,contacts);
    listView1.setAdapter(adapter);
    // Register the ListView  for Context menu
    registerForContextMenu(listView1);
  }
  @Override
  public void onCreateContextMenu(ContextMenu menu, View v, ContextMenuInfo menuInfo)
  {
      super.onCreateContextMenu(menu, v, menuInfo);
      menu.setHeaderTitle("Select The Action");
      menu.add(0, v.getId(), 0, "Call");//groupId, itemId, order, title
      menu.add(0, v.getId(), 0, "SMS");
  }
  @Override
  public boolean onContextItemSelected(MenuItem item){
      if(item.getTitle()=="Call"){
        Toast.makeText(getApplicationContext(),"calling code",Toast.LENGTH_LONG).show();
      }
      else if(item.getTitle()=="SMS"){
        Toast.makeText(getApplicationContext(),"sending sms code",Toast.LENGTH_LONG).show();
      }else{
        return false;
      }
    return true;
  }
}
```

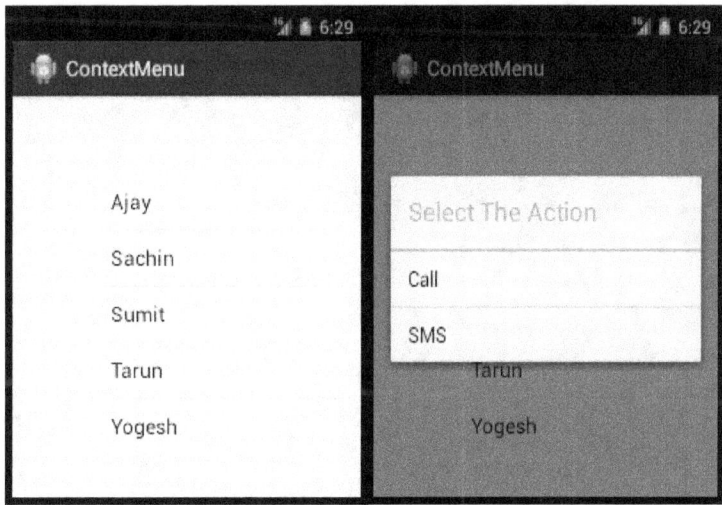

Figure 6.16: Context Menu - Android

www.ingramcontent.com/pod-product-compliance
Lightning Source LLC
Chambersburg PA
CBHW082106220526
45472CB00009B/2070